Science of the Cosmos, Science of the Soul

Science of the Cosmos, Science of the Soul

The Pertinence of Islamic Cosmology in the Modern World

William C. Chittick

ONEWORLD

SCIENCE OF THE COSMOS, SCIENCE OF THE SOUL

A Oneworld Book
Published by Oneworld Publications 2007

ACIP record for this title is available
from the British Library

ISBN 978–1–85168–495–3

Typeset by Jayvee, Trivandrum, India
Cover design by e-Digital Design
Printed and bound in Great Britain
by Lightning Source

Oneworld Publications
10 Bloomsbury St
London WC1B 3SR
England

Contents

Introduction vii

1. **A Vanishing Heritage** 1

Ijtihād 3
The Role of the Intellectual Tradition 5
The Current Situation 8
The Gods of Modernity 12
The Goal of Intellectual Understanding 16
The Rejection of Tradition 18

2. **Intellectual Knowledge** 23

Verifiable Knowledge 25
Intellect 29
Basic Findings 30
A Visitor from the Past 33

3. **The Rehabilitation of Thought** 39

Thought 40
The Intellectual Tradition 44
Taqlīd and *Taḥqīq* 45
Premodern Science 47
The Reign of *Takthīr* 52
The Goal of Thought 55

4. Beyond Ideology 59

The Omnipresence of Transmission 60
Breaking the Shell of Dogmatism 62
Asserting Absoluteness 65
Mythic Imagination 69
Self-Understanding 73

5. The Unseen Men 75

Sufism 79
Cosmos and Soul 82
Naming Reality 84
The One and the Many 90
The Living Universe 92
Islamic Science 94
The Efficacy of Names 96
Inadequate Names 99
The Myth of the Unseen Men 101

6. The Anthropocosmic Vision 109

Ahistorical and Historical Knowledge 110
The Philosophical Quest 114
The Methodology of *Taḥqīq* 118
Understanding the Soul 121
Origin and Return 125
Omniscience 128

7. The Search for Meaning 133

Two Modes of Knowing 134
Subject and Object 136
The Worldview 138
The Self 142
Meaning 144

Notes 151
Index 153

Introduction

I began studying Islamic thought forty years ago. I was originally attracted to the field by a fortuitous set of circumstances that led me to spend my junior year in college at the American University of Beirut. A general interest in non-Western religions blossomed when I was exposed to lectures and books on Sufism and Islamic philosophy. I quickly realized that the only way to acquire more than a superficial acquaintance with these topics was to learn Arabic and Persian. After a dozen years of study and research, I began publishing the results of my explorations. My primary concern from the beginning was trying to understand what Sufis and Muslim philosophers were saying. How did reality appear to them? How did they explain the great issues of meaning that people face in attempting to make sense of their lives?

In most of my publications over the years, I have let Rūmī, Ibn 'Arabī, Ṣadr al-Dīn Qūnawī, 'Abd al-Raḥmān Jāmī, Afḍal al-Dīn Kāshānī, Shams-i Tabrīzī, Mullā Ṣadrā, and others do the talking, while I sat back with my readers and listened to their words. In the past few years, however, I have felt more at ease in applying the insights gleaned from the material to new contexts. Given the deep seriousness of the authors, it has seemed to me that

I owe it to them to bring out some of the significance of their perspectives for the specifically modern context, such as the role played by science in the contemporary *Zeitgeist*. It is the attempt to find contemporary relevance that is the common thread of these essays.

Much of the book develops implications of a distinction between two ways of knowing that is basic to the great religions under a variety of nomenclature, though it is typically ignored in discussions of contemporary issues. Islamic sources speak about it in a variety of ways. Here I focus on a standard differentiation that is made between "transmitted" (*naqlī*) and "intellectual" (*ʿaqlī*).

Transmitted knowledge is characterized by the fact that it needs to be passed from generation to generation. The only possible way to learn it is to receive it from someone else. In contrast, intellectual knowledge cannot be passed on, even though teachers are needed for guidance in the right direction. The way to achieve it is to find it within oneself, by training the mind or, as many of the texts put it, "polishing the heart." Without uncovering such knowledge through self-discovery, one will depend on others in everything one knows.

Typical examples of sciences based on transmitted learning are language, history, and law. The usual example of an intellectual science, even though it does not meet all the criteria, is mathematics. We do not say, "Two plus two equals four because the authorities say so." The mind is able to discover and understand mathematical truth on its own, and once it discovers it, it does not depend on outside sources. The knowledge is known to be true because, once we understand it, it is self-evident. We can no more deny its truth than we can deny our own awareness.

Transmitted knowledge depends on hearsay. It is by far the most common sort of knowledge in any culture or religion. Buddhists may know that enlightenment is an experience that transcends all conventional forms of knowing, but, until they achieve it, they have received what they know about it by way of transmission. Muslims know that God requires them to pray five times a day, but they take this knowledge from the ulama, those who have become learned in the Qur'an and the Hadith. They cannot discover what God wants from them without the transmission of the revealed sources. So also for the rest of us: transmission and hearsay provide us with language, culture, opinions, worldview, and practically everything we think we know. In contrast, intellectual understanding is what we know with complete certainty in the depths of our souls. But such knowledge is rare.

The search for intellectual knowledge in Islamic civilization was undertaken in two broad fields of learning, each of which developed many branches and underwent numerous historical vicissitudes. For simplicity's sake, I am calling them philosophy and Sufism. Philosophy built on the logical and rational methodologies systematized by the Greeks, and Sufism based itself on the contemplative techniques received from the Prophet. The two fields frequently overlapped, especially from the thirteenth century onward.

Philosophy and Sufism diverged sharply from the transmitted sciences by acknowledging explicitly that the meanings of things in the world cannot be found without simultaneously finding the meaning of the self that knows. Certainly, one studies the world to achieve the understanding of phenomena, but understanding is an attribute of the soul, of the knowing subject. Masters of the intellectual approach recognized that meaning hides behind the "signs" (*āyāt*) of God, that all phenomena point to noumena, and that those noumena can only be accessed at the root of the knowing self.

If we view the intellectual tradition in a broad perspective, it is clear that it did not allow for the sharp distinction between subject and object that was a prerequisite for the rise of modern science. If I focus more on Islamic philosophy than on Sufism here, it is partly because of the notion often seen in the writings of Western historians and modern-day Muslim apologists that Islamic science – which was developed by the philosophers and not the Sufis – was an important precursor to modern science. I chose the title "Science of the Cosmos, Science of the Soul" precisely because it highlights science and at the same time brings in the term "soul," which is central to the philosophical tradition and about as unscientific as a term can be.

Let me say up front that the intellectual approach about which I am writing has been moribund for over a century. A few people still speak for it, but their voices go largely unheard. The economic, political, and social forces that drive activity in the rest of the world have not left Muslims behind. Those who are able to gain an education normally do so with pecuniary goals in mind. The technical and practical fields, which can be mastered rather quickly and offer relative assurance of a comfortable life, attract the best students and dominate the universities. The traditional educational institutions, which used to ask students to dedicate their lives to the quest for knowledge and virtue, have almost totally disappeared. In their places have grown up "theological" schools that churn out zealots and ideologues.

The first four chapters address the disappearance of the intellectual tradition and the numerous obstacles that stand in the way of its recovery. Chapter One provides a brief explanation of the nature of this tradition and describes various forces, both internal and external to the Muslim community, that have obscured its importance. Chapter Two expands on the distinction between transmitted and intellectual learning, discusses basic elements of the philosophical and Sufi worldview, and tries to suggest the oddity of our own historical situation by looking at ourselves through the eyes of an imagined Muslim intellectual. Chapter Three continues the discussion of obstacles to recovery and ways to overcome them. Chapter Four looks at ideology as a pillar of modern thought and suggests how the intellectual tradition might help people break its spell.

The final three chapters look more carefully at the actual teachings of the intellectual tradition, focusing on their relevance to contemporary questions of science and meaning. Chapter Five reflects on the philosophy of Seyyed Hossein Nasr, one of the few today who speak for this tradition, and it looks at the centrality of language in human nature and the manner in which the mythic imagination structures the interrelationship between cosmos and soul through the process of naming. Chapter Six attempts to explain why the philosophical worldview refused to differentiate sharply between subject and object, and how the quest for self-knowledge provides the key to the profound difference between the Islamic and the modern understandings of "science." Chapter Seven focuses on the quest of aspiring intellectuals to transcend egocentricity and specifying objectives so as to achieve freedom from all constraints.

Except the fifth, all the chapters were originally written as lectures, and most have been published in that form. All have been thoroughly revised if not totally rewritten with a view toward integration. The first three were delivered to Muslim audiences, which helps explain their sharper focus on Islamic concepts and rhetoric. The other chapters were written for more general audiences, so I have avoided some of the specifically Islamic notions and brought in references to other religious and intellectual traditions.

William C. Chittick
Stony Brook University

1

A Vanishing Heritage

Intellectual understanding in the strict sense is found at the highest pinnacle of human selfhood, what the philosophers call the "actual intellect." When such understanding leaves the realm of pure intelligence and descends to the level of thought and language, we are dealing with its expression, which will always be inadequate. To begin with, expression is simply transmitted knowledge, not actual understanding. Nonetheless, we can still appreciate that a distinction has always been drawn between these two sorts of knowledge in Islam and other traditions. It is this distinction that I need to clarify at the outset. Then I will suggest how ignorance of the foundational importance of intellectual understanding has contributed to the crises faced not only by Muslims, but also by the human community in general.

The intellectual tradition in Islam has addressed four basic topics: God, the cosmos, the human soul, and interpersonal relationships. The first three are foundational constituents of reality as we perceive it, and the fourth applies the insights gained from studying the first three to the realm of human activity. One can of course read about all these topics in the authoritative sources of transmitted knowledge, such as the Qur'an and the Hadith, but knowing them for oneself is another matter altogether. For the intellectual tradition, transmitted knowledge plays the role of pointers toward an understanding that must be actualized and realized by the seeker.

Perhaps the best way to understand the difference between transmitted and intellectual knowledge is to reflect upon the difference between "imitation" or "following authority" (*taqlīd*) and "realization" or "verification" (*taḥqīq*), terms that designate the two basic paths of acquiring knowledge. In order to be a member of any religion, culture, society, or group, one needs to learn from those who are already members, and this process of learning goes on by way of "imitation." This is how we learn language and culture, not to mention scripture, ritual, and law. In the Islamic context, those who have assumed the responsibility of preserving this transmitted heritage are called the ulama, that is, the "knowers" of the tradition.

In transmitted knowledge, the question of "why" is pushed into the background. When someone asks the ulama why one must accept such-and-such a dogma or why one must pray or fast, the basic answer is "because God said so," which is to say that we have the knowledge on the authority of the Qur'an and the Sunnah. In the same way, parents correct their children's speech by calling on the authority of usage or the rules of grammar.

Intellectual knowledge is altogether different. If one accepts it on the basis of hearsay, one has not understood it. Mathematics is a science that does not depend on the authorities. Rather, it needs to be awakened in one's awareness. In learning it, students must understand why, or else they will simply be imitating others. It makes no sense to say that two plus two equals four because my teacher said so. Either you understand it, or you don't. You must discover its truth within yourself. The Muslim intellectuals held that to imitate others in *intellectual* issues is the status of a beginner or a student, not a master, but to imitate the Qur'an and the Prophet in *transmitted* matters is to follow the right path.

In short, there are two basic sorts of knowledge, and each has methods proper to it. *Taqlīd* or imitation is proper to the transmitted sciences, and *taḥqīq* or realization is proper to the intellectual sciences.

IJTIHĀD

The word *taqlīd* is often discussed in the writings of modern-day Muslim thinkers, who typically describe it as the bane of Islamic society. These discussions, however, do not focus on *taqlīd* as the opposite of *taḥqīq*, but rather as the opposite of *ijtihād*. Given the prominence of this issue among contemporary Muslim writers, I need to make clear at the outset that I am talking about something else.

Ijtihād means the achievement of sufficient mastery in the discipline of jurisprudence (*fiqh*) to exercise independent judgment in deriving the Shariah (Islamic law). Someone who reaches this rank is called a *mujtahid*. Such a person does not need to follow the authority of other jurists in matters of the Shariah. Nonetheless, his or her mastery remains on the level of transmitted knowledge, which is to say that it is still based on the Qur'an, the Hadith, and reports from the forefathers and the masters of the discipline. Given the qualifications needed to become a *mujtahid*, most Sunni Muslims over the past few centuries have held that the gate of *ijtihād* is closed. Shi'ites, in contrast, consider it always open.

From the point of view of jurisprudence, a person who is not himself a *mujtahid* must imitate someone who is – whether the *mujtahid* be alive (as in Shi'ism) or long dead (as in Sunnism). One follows a *mujtahid* because one can only learn the Shariah from someone who already knows it. This is not the situation in the intellectual sciences, however. A *mujtahid*, with all his or her mastery of the transmitted science of jurisprudence, is by no means a *muḥaqqiq*, one who has achieved *taḥqīq* or realization in intellectual knowledge. To begin with, intellectual knowledge does not depend upon transmission. A *muḥaqqiq* can, in principle, grasp all the intellectual sciences without the help of past generations or divine revelation. You do not need a prophet to tell you that two plus two equals four or that God is one. The knowledge itself, once known, is self-evident, which is to say that it carries its own proof in the very act of understanding it.

The ulama of the Shariah implicitly recognize the differing nature of intellectual knowledge when they tell us, as they often do, that faith (*īmān*) on the basis of imitation is unacceptable to God. A Muslim cannot be true to his tradition if he says, "I have faith in God because my parents told me to." Someone like this would be saying that if he had been told not to believe in God, he would not, so his faith would be empty words.

Although in theory we can distinguish between transmitted and intellectual knowledge, in practice the two have always been closely intertwined, and the intellectual sciences have always built on the transmitted sciences. One cannot speak properly without grammar, and one cannot understand specifically Islamic teachings without the Qur'an and the Hadith. However, the fact that people may have an excellent knowledge of the transmitted sciences does not mean that they know anything at all about the intellectual sciences. Nor does the ability to recount the metaphysical and cosmological theories of the great Muslim intellectuals prove that a person has any understanding of what the theories mean.

Both the transmitted and the intellectual sciences are essential to the survival of any religion, but both are now being lost. By and large, however, transmitted sciences are better preserved than intellectual sciences, and the reason is obvious. Anyone can memorize Qur'an and Hadith, but few can truly understand what God and the Prophet are talking about. One can only understand in one's own measure, and fewer and fewer people undertake the training necessary to develop their talents and capacities.

It is obvious that one cannot understand mathematics (or any other science) without both native ability and training. Even if one happens to have a great aptitude, one will never get very far without years of study. If this is true of sciences like mathematics or grammar, which deal with realities that are relatively near at hand, it is much more true of metaphysics, which deals with the deepest realities, the furthest from our everyday experience.

THE ROLE OF THE INTELLECTUAL TRADITION

It is important to stress that no religion can survive, much less flourish, without a living intellectual tradition. This becomes clear as soon as we ask ourselves the questions: What was the intellectual tradition for? What function did it play in society? What was its goal? In other words: Why should people think? Why shouldn't they just blindly accept whatever they're told? The basic Muslim answer is that people should think because they must think, because they are thinking beings. They have no choice but to think, because God has given them minds and intelligence. Not only that, but in numerous Qur'anic verses God has commanded them to think and to employ their intelligence. To think properly a person must actually think, which is to say that conclusions must be reached by one's own intellectual struggle, not by someone else's. Any experienced teacher knows this perfectly well.

No doubt, this does not mean that God requires everyone to enter into the sophisticated sort of study and reflection that went on in the intellectual sciences, because not everyone has the requisite talents, capacities, and circumstances. Nonetheless, people have the moral and religious obligation to use the minds that God has given them. As the Qur'an puts it, "God burdens a soul only to its capacity" (2:286). If people's capacity includes thinking, then they have the duty to think. But God does not tell them *what* to think, because that would be to make imitation and hearsay incumbent in intellectual matters. It would defeat the very purpose of human intelligence, which is for us to understand for ourselves.

No doubt many if not most people are unreflective and never even ask themselves why they should bother thinking about things. They simply go about their daily routine and imagine that they understand their situation. Observant Muslims of this sort seem to assume that God wants nothing more from them than following the Shariah. But this is no argument for those who have the ability to think. Anyone who has the capacity and talent to reflect upon God, the universe, and the human soul has the duty to do so. Not to do so is to betray one's own nature and to disobey God's instructions to ponder the signs.

Given that some Muslims have no choice but to think, learning how to think correctly must be an important area of Muslim effort. But what defines "correct" thinking? How do we tell the difference between right thinking and wrong thinking? Does the fact that people have no choice but to think mean that they are free to think whatever they want? The Islamic answer to these sorts of question has always been that the way people think is far from indifferent. Some modes of thinking are encouraged by the Qur'an and the Sunnah, and some are discouraged. Islamically, it is incumbent upon those who think to employ their minds in ways that coincide with the goals of the Qur'an and the Sunnah. In other words, the goal of the Islamic intellectual tradition needs to coincide with the goal of Islam, or else it is not Islamic intellectuality.

So, what is the goal of Islam? In general terms, Islam's goal is to bring people into harmony with the way things actually are. In other words, it is to bring them back into the presence of God, from which they emerged in the first place. However, everyone is going back into God's presence in any case, so the issue is not going back per se, but how one gets there. Through the Qur'an and the Sunnah, God guides people back to him in a manner that will ensure their permanent happiness. If they want to follow a "straight path" (*ṣirāṭ mustaqīm*), one that will lead to balance and happiness and not to disequilibrium and misery, they need to employ their minds, awareness, and thinking in ways that harmonize with God himself, the true Reality. If they occupy themselves with illusion and unreality, they will follow a crooked path and most likely not end up where they would like to go.

The history of Islamic intellectual expression is embodied in the various forms that Muslims have adopted over time in attempting to think rightly and correctly. The intellectual tradition was robust and lively, so disagreements about the best way to express its findings were common. Nonetheless, in all the different schools of thought that have appeared over Islamic history – whether they dealt with intellectual or transmitted learning – one principle has always been agreed upon: God is one, and he is the only source of truth and reality. He is the origin of all things, and all things return to him. This is *tawḥīd*, "asserting the unity of God." It is expressed most succinctly in the first half of the Shahadah, the testimony of faith: "There is no god but God." This statement is commonly known as *kalimat al-tawḥīd*, "the words asserting unity." To think Islamically is to recognize God's unity and to draw the proper consequences. Differences of opinion arise concerning the proper consequences, not the fact that God is one.

The consequences that people draw from *tawḥīd* depend largely on their understanding of God. Typically, Muslims have sought to understand God by meditating on the implications of God's names and attributes as expressed in the Qur'an and the Sunnah. If God is understood as a lawgiver, people will draw conclusions having to do with the proper observance of the Shariah. If he is understood as wrathful, they will conclude that they must avoid his wrath. If he is understood as merciful, they will think that they must seek out his mercy. If he is understood as beautiful, they will find him lovable.

God, of course, has "ninety-nine names" – at least – and every name throws a different light on what exactly God is, what exactly he is not, and how exactly people should understand him and relate to him. Naturally, thoughtful Muslims have always understood God in many ways, and they have drawn diverse conclusions on the basis of each way of understanding. This diversity of understanding in the midst of *tawḥīd* is prefigured in the Prophet's prayer, "O God, I seek refuge in Your mercy from Your wrath, I seek refuge in Your good pleasure from Your anger, I seek refuge in You from You."

THE CURRENT SITUATION

I said that the Islamic intellectual tradition has largely, though not completely, disappeared. This is obvious to those who have studied the history of Islamic civilization. Scholars often discuss its disappearance in terms of the "golden age" of classical Islam and the gradual decline of science and learning. Given that almost everyone agrees that Islamic scholarship in its various forms does not match up to its greatness in the past, there is little to be gained by trying to prove the point, or by mapping out the history of the decline, or by suggesting what may or may not have gone wrong.

Instead, I want to assume that the intellectual tradition is not what it used to be, and that it still has something to offer. What this something is, however, cannot be rediscovered or resuscitated as long as intellectual knowledge is treated as another form of transmitted knowledge, as is normally done by modern scholarship. We have numerous experts in Islamic philosophy and Sufism, among both Muslims and non-Muslims, who have made tremendous contributions to textual and historical studies but who deal with their subject as a repository of historical information, not as a living tradition whose *raison d'être* is the transformation of the human soul. As one of my old and now deceased professors at Tehran University once remarked about his young colleagues, they know everything one can possibly know about a text, except what it says.

Despite the definition of "intellectual" provided earlier, some people will claim that the Muslim community has a vibrant intellectual life and that the intellectual tradition is not in fact disappearing. But this would be to fall back on current meanings of the word intellectual. No doubt there are tens of thousands of Muslim intellectuals in the ordinary sense – that is, writers, professors, doctors, lawyers, and scientists who are concerned with current affairs and express themselves vocally or in writing. But I have serious doubts as to whether any more than a tiny fraction of such people are intellectuals in the technical sense in which I am defining the term. Yes, there are many thoughtful and sophisticated people who were born as Muslims and who may indeed practice their religion carefully. But do they think Islamically? Is it possible to be both a scientist in the modern sense and a Muslim who understands the cosmos and the soul as the Qur'an and the Sunnah explain them? Is it possible to be a sociologist and at the same time to think in terms of *tawḥīd*?

As soon as we have an idea of the nature of the intellectual tradition in the sense of the word that I have in mind, then it will appear highly likely that the thought processes of most Muslim thinkers today are not in fact determined by Islamic principles and Islamic understanding. Rather, they are shaped and molded by habits of mind learned unconsciously in grammar and high school and then confirmed and solidified by university and professional training. Such people may pray and fast like Muslims, but they think like doctors, engineers, sociologists, and political scientists.

It is naive to imagine that one can learn how to think Islamically simply by attending lectures once a week or by reading a few books written by contemporary Muslim leaders, or by studying the Qur'an, or by saying one's prayers and having "firm faith." In the premodern Islamic world, the Islamic ethos was everywhere, but great thinkers and intellectuals still spent their whole lives searching for deeper knowledge of God, the cosmos, and the soul. As far as they were concerned, the search for understanding was a never-ending task.

The heritage left by those intellectuals is extraordinarily rich. They wrote many thousands of books, even if most of the important books have never been printed, much less translated. And, those that have been published are rarely read by modern-day Muslims. I do not mean to imply that it would be necessary to read all the great books of the intellectual tradition in their original languages in order to think Islamically. If modern-day Muslims could read one of the important books, even in translation, and understand it, their thinking would be deeply affected. However, the only way to understand such books is to prepare oneself for understanding, and that demands study and training. This cannot be done on the basis of a modern university education, unless, perhaps, one has devoted it to the Islamic tradition (I say "perhaps" because many Muslims and non-Muslims with Ph.D.'s in Islamic Studies cannot read and understand the great books of the intellectual heritage).

Given that modern schooling is rooted in topics and modes of thought that are not harmonious with traditional Islamic learning, it is profoundly difficult for any thinking and practicing Muslim to harmonize the domain of thought and theory with the realm of faith and practice. One cannot study for many years and then be untouched by what one has studied. There is no escape from picking up mental habits from the material to which one devotes one's life. It is most likely, and almost, but not quite, inevitable for modern thinkers with religious faith to have compartmentalized minds. One compartment of the mind will encompass the professional, rational domain, and the other the domain of personal piety and practice. More generally, this is the case with most people who grew up in a traditional ambience and were then educated in the modern style. The Iranian thinker Daryoush Shayegan, who writes eloquently as a philosopher and social critic while expressing his own personal struggle with this phenomenon, calls it "cultural schizophrenia."[1]

Believers of a thoughtful bent who are caught up in cultural schizophrenia may try to rationalize the relationship between their religious practice and their professional training, but they will do so in terms of the worldview determined by the rational side of the mind. The traditional Islamic worldview, established by the Qur'an and passed down by generations of Muslims, will be closed to them, and hence they will draw their categories and ways of thinking from the ever-shifting *Zeitgeist* that is embodied in contemporary cultural trends and popularized through television and other forms of mass indoctrination.

Many Muslim scientists tell us that modern science helps them see the wonders of God's creation, and this is certainly an argument for preferring the natural sciences over the social sciences. But is it necessary to study physics or biochemistry to see the signs of God in all his creatures? The Qur'an keeps on telling Muslims, "Will you not reflect, will you not ponder, will you not think?" About what? About the signs, which are found, as over two hundred Qur'anic verses remind us, in everything, especially natural phenomena. It does not take a great scientist, or any scientist at all, to understand that the world speaks loudly of the majesty of its Creator. Any fool knows this. This is what the Prophet called the "religion of old women" (*dīn al-'ajā'iz*), and no one needs any professional training to understand it.

It is true that a basic understanding of the signs of God may provide sufficient knowledge for salvation. After all, the Prophet said, "Most people of paradise are fools." However, the foolishness that leads to paradise demands foolishness in the affairs of this world, and nowadays that is not easy to come by. It is certainly not often found among Muslim thinkers, who are already far too clever, which helps explain why they are such successful doctors and engineers.

THE GODS OF MODERNITY

Perhaps the best way to demonstrate that the habits of mind imparted by modernity are seldom congruent with Islamic learning is to reflect on the characteristics of modernity – by which I mean the thinking and norms of the "global culture" in which we live today. It should be obvious that whatever characterizes modernity, it is not *tawḥīd*, the first principle of Islamic thinking. Rather, it is fair to say that modernity is characterized by the opposite of *tawḥīd*. One could call this *shirk* or "associating others with God," but for most Muslims, this word is too emotionally charged to be of much help in the discussion. So, let me call the characteristic trait of modernity *takthīr*, which is the literal opposite of *tawḥīd*. *Tawḥīd* means to make things one, and, in the religious context, it means asserting that God is one. *Takthīr* means to make things many, and, as I understand it here, it means asserting that the gods are many.

Modern times and modern thought lack a single center, a single orientation, a single goal, any single purpose at all. In other words, there is no single "god." A god is what gives meaning and orientation to life, and the modern world derives meaning from many, many gods. Through an ever-intensifying process of *takthīr*, the gods have been multiplied beyond count, and people worship whatever gods appeal to them.

The process of increasing *takthīr* becomes clear when we compare the general course of Islamic thought over history with that of European civilization. Up until recent times, Islamic thought was characterized by a tendency toward unity, harmony, integration, and synthesis. The great Muslim thinkers were masters of many disciplines, but they looked upon them as branches of the single tree of *tawḥīd*. There was never any contradiction between astronomy and zoology, or physics and ethics, or mathematics and law, or mysticism and logic. Everything was governed by the same principles, because everything fell under God's all-encompassing reality.

The history of European thought is characterized by the opposite trend. Although there was a great deal of unitarian thinking in the medieval period, from that time onward dispersion and multiplicity have constantly increased. "Renaissance men" could know a great deal about all the sciences and at the same time have a unifying vision. But nowadays, everyone is an expert in some tiny field of specialization, and information increases exponentially. The result is mutual incomprehension and universal disharmony. It is impossible to establish any unity of understanding, and no real communication takes place among specialists in different disciplines. Since people have no unifying principles, the result is an ever-increasing multiplicity of goals and gods, an ever-intensifying chaos.

Everyone worships some god or another. No one can survive in an absolute vacuum, with no goal, no significance, no meaning, no orientation. The gods that people worship are those points of reference that give meaning and context to their lives. The difference between traditional objects of worship and modern objects of worship is that in modernity, it is almost impossible to subordinate all the minor gods to a supreme god, and, when this is done, the supreme god has been manufactured by ideologies. It is certainly not the God of *tawḥīd*, who is the absolute and supreme reality, next to whom nothing else is real. However, it may well be an imitation of the God of *tawḥīd*, especially when religion enters the field of politics.

The gods in a world of *takthīr* are legion. To mention some of the more important ones would be to list the defining myths and ideologies of our times – freedom, equality, evolution, progress, science, medicine, nationalism, socialism, democracy, Marxism. But perhaps the most dangerous of the gods are those that are the most difficult to recognize. They have innocuous names like care, communication, consumption, development, education, information, standard of living, management, model, planning, production, project, resource, service, system, welfare.

Those who do not think that these words play the role of gods should take a look at *Plastic Words* by Uwe Poerksen. The book's subtitle is more instructive: *The Tyranny of a Modular Language.* Poerksen explains that the modern use of language – a use that achieved dominance after World War II – has produced a group of words that have turned into the most destructive tyrants the world has ever seen. He does not call them "gods," for he writes as a linguist and has no apparent interest in theology. Nonetheless, he does give them the label "tyrant," and this is a good translation for the Qur'anic divine name *jabbār*. When this name is applied to God, it means that God has absolute controlling power over creation. "Tyranny" becomes a bad thing when it is claimed by creatures, for it indicates that they have tried to usurp God's power and authority. In the case of the plastic words, power has been usurped by words that shape discussion of societal goals.

As Poerksen points out, these tyrannical words have at least thirty common characteristics. The most important is that they have no definition, though they do have an aura of goodness and beneficence about them. In linguistic terms, this is to say that they have many connotations but no denotation. There is no such thing as "care" or "welfare" or "standard of living," but the words suggest many good things to most people. They are abstract terms that seem to be scientific, so they carry an aura of authority in a world in which science is one of the most important gods. Each of them turns something indefinable into a limitless ideal and awakens endless needs. Once the needs are awakened, they seem to be self-evident and quickly turn into necessities. The Qur'an says that God is rich, and people are poor and needy toward God. Nowadays, people feel poor and needy toward these little tyrants.

Those who speak on behalf of the plastic words gain power and prestige, for they represent science, freedom, and progress. As a result, dissenting voices are ignored and marginalized, since, we imagine, only a complete idiot would object to care and development. Everyone must follow those whose only concern is to care for us and to help us develop.

The ulama who speak for these mini-gods are the "experts". Each of the plastic words sets up an ideal and encourages us to think that only the experts can show us how to achieve it, so we must entrust our lives to them. We must imitate the scientific ulama, who lay down shariahs for our health, welfare, and education. People treat the pronouncements of experts as fatwas (legally binding opinions on points of law). If the experts reach consensus (*ijmāʿ*) that we must destroy a community as a sacrificial offering to development, then we have no choice but to follow their authority. The ulama know best.

Each of the plastic words makes other words appear backward and out-of-date. We can be proud of worshiping these gods, and all of our friends and colleagues will consider us enlightened whenever we recite the proper litanies in praise of them. Those who still take the old God seriously can hide this embarrassing fact by worshiping the new gods along with the old. And obviously, many people who continue to claim that they are worshiping the old God will twist his teachings so that he also seems to be telling us to serve the new gods.

THE GOAL OF INTELLECTUAL UNDERSTANDING

Understanding the nature of false gods has always been central to the intellectual sciences, but it cannot be the concern of the transmitted sciences. One cannot accept *tawḥīd* simply on the basis of imitation, which is to say that it stands outside the domain of transmitted learning. *Tawḥīd* must be understood if people are to have faith in it, even if their understanding is far from perfect. Much of the intellectual tradition has been concerned with explaining *tawḥīd* and the manner in which it clarifies the objects of faith – God, angels, scriptures, prophets, the Last Day, and "the measuring out, the good of it and the evil of it." How are Muslims to understand these objects? Why should they have faith in them? True faith can never be blind belief, but rather commitment to what one actually knows to be true.

In discussing God and the other objects of faith in the light of *tawḥīd*, it is important to explain not only what they are, but also what they are not. When people do not know what God is, it is easy for them to fall into the habit of worshiping false gods, and that leaves them with no protection against the *takthīr* of the modern world, the multiplicity of gods that modern ways of thinking demand that they serve.

What is striking about contemporary Islam's encounter with modernity is that Muslims lack the intellectual preparation to deal with the situation. Muslim thinkers – with a few honorable exceptions – do not question the legitimacy of the modern gods.

Rather, they debate over the best way to serve them. In other words, they think that Islamic society must be modified and adapted to achieve the ideals represented by the gods of modernity, and especially those designated by the plastic words. This is to say that innumerable modern-day Muslims are forever looking for the best ways to bring their society into conformity with the rejection of *tawḥīd*.

Many Muslims today recognize that the West has paid too high a price for modernization and secularization. They see that various social crises have arisen in all modernized societies, and they understand that these crises are somehow connected with the loss of religious traditions, the ultimate meaninglessness of modern life, and the devaluation of ethical and moral guidelines. But many of these same people tell us that Islam is different. Islam can adopt the technology and the know-how – the progress, development, and expertise – while preserving its own moral and spiritual strength and avoiding the social disintegration of the West. In other words, they think, Muslims can forget *tawḥīd*, embark on a course of *takthīr*, and suffer no negative consequences.

Especially surprising here is the extent to which contemporary Muslims seem to think that an Islamic order can be imposed by modern states, with their historically unprecedented ability to indoctrinate and coerce. The actual attempts to do so demonstrate clearly that an "Islamic" society can easily be turned into another version of the monstrous totalitarianisms that have been all too characteristic of the modern world. The pervasiveness of bureaucracy, technology, and the worldview of *takthīr* and their steady encroachment on all human relationships mean that more and more of the world is dehumanized, reified, and opened up to manipulation. Traditional moral constraints carry little weight in face of the institutions of modernity, especially at the time of crisis – and when has there not been a crisis?

The fact that so many people think that Islam can flourish and simultaneously adopt the gods of modernity shows that they have lost the vision of *tawḥīd* that used to give life to Islamic thinking. They cannot see that everything is interrelated, and they fail to understand that the worship of false gods necessarily entails the dissolution of every sort of order – the corruption not only of individuals and society, but also of the natural world. In other words, when people refuse to serve God as reality itself demands that they serve him, they cannot fulfill their human functions. When people refuse to live in harmony with the transcendent principles that determine the way things actually are, they bring about chaos and disorder in the natural and social environments. The Qur'an sums up the process in the verse, "Corruption has appeared in the land and the sea because of what the hands of people have earned" (30:41). "Corruption" (*fasād*) is defined as the lack of "wholesomeness" (*ṣalāḥ*), and wholesomeness is wholeness, health, balance, harmony, coherence, order, integration, and unity on the individual, social, and cosmic levels. It can be established only through *tawḥīd* or "making things one."

THE REJECTION OF TRADITION

Major obstacles prevent the recovery of the intellectual heritage. These can be discerned on the societal level in the diverse beliefs and attitudes that have been adopted by modern-day Muslims as a result of their loss of intellectual independence and their blind imitation of the norms embodied in the ideals, institutions, and structures of the modern world. Among these obstacles are politicization of the community, monolithic interpretations of Islamic teachings, and unthinking acquiescence to the ideological preaching of Muslim leaders. Perhaps the deepest and most pernicious of these obstacles, however, is the general trend to reject all but the most superficial trappings of the Islamic tradition.

Like other religions, Islam is built on tradition, by which I mean the sum total of the transmitted and intellectual heritages. Nonetheless, many Muslims see no contradiction between believing in the gods of modernity and accepting the authority of the Qur'an and the Sunnah. In order to do this, they need to ignore thirteen hundred years of Islamic intellectual history and pretend that no one needs the help of the great thinkers of the past to understand and interpret the Qur'an and the Sunnah.

We need to keep in mind that the only universally accepted dogma in the modern world is the rejection of tradition. The great prophets of modernity – Descartes, Rousseau, Marx, Freud – followed a variety of gods, but they all agreed that the old gods were no longer of any use. In the Islamic view, God's prophets share *tawḥīd*. The prophets of modernity share *takthīr*. One can only reject God's unity by inventing other gods to replace him.

In Islamic theology, God is *qadīm*, "ancient" or "eternal." He has always been and always will be. In modernity, the gods are new. To stay new, they have to be changed or modified frequently. The new is always to be preferred over the old, which is "outmoded" and "backward." Science is always making new discoveries, and technology is constantly offering new inventions that quickly become necessities. Anything that is not in the process of renewal is thought to be dead.

One name for this god of newness is "originality." He rules by ordaining new styles and models, and his priests are found everywhere, especially in advertising and mass indoctrination. The fashion *mujtahids* tell women what to wear, but they change their fatwas every year. The world of art blatantly and openly worships Originality as the highest god. Or take the modern university, where professors often adopt the latest theories as soon as they arrive from Paris.

The greatest danger of the hostility toward tradition that is so common among modern-day Muslims is that they have accepted the god of newness – like so many others – without giving any thought to what they are doing. As far as they are concerned, Muslim thinkers and intellectuals have had nothing to say for thirteen hundred years. They would like to retain their Muslim identity, but they imagine that in order to do this, it is sufficient to keep their allegiance to the Qur'an and the Sunnah and ignore its great interpreters.

For such people, the ruling gods are progress, science, and development. They imagine that we know so much more about the world than those people of olden times, because "we" have science. Of course, they themselves do not have science, they have simply heard and believed that scientific knowledge is real knowledge. They know little about the goals and methods of science, and nothing about the Islamic intellectual tradition. They are blind imitators in intellectual issues, that is, on the level where they should be striving for their own understanding. What is worse, this is a selective imitation, since they only accept the authority of the "scientists" and the "experts," not that of the great Muslim thinkers of the past. If Einstein said it, it must be true, but if al-Ghazālī or Mullā Ṣadrā said it, then it can't be true, because it isn't scientific.

Finally, let me suggest that the most basic problem of modern Islam, a problem present in every religion, is that believers suffer from what has traditionally been called "compound ignorance" (*jahl murakkab*). "Ignorance" is not to know. "Compound ignorance" is not to know that you do not know. Too many Muslims do not know what the Islamic tradition is, they do not know how to think Islamically, and they do not know that they do not know. The first step in curing ignorance is to recognize that you do not know. Once people recognize their own ignorance, they can go off in "search of knowledge" (*ṭalab al-ʿilm*) – a search which, as the

Prophet said, "is incumbent on every Muslim," and indeed, one would think, on every human being.

No recovery of the intellectual tradition will be possible until individuals take steps for themselves. The tradition can never be recovered by imitation or by community action, only by individual dedication and personal realization. Governments and committees cannot begin to solve the problem. Understanding cannot be imposed or legislated, it can only grow up in the heart.

2

Intellectual Knowledge

If we remember nothing else about intellectual knowledge, we need to keep in mind that it is achieved by *taḥqīq*, which is to know things by verifying and realizing their truth and reality for oneself. One cannot verify the truth and reality of things without knowing them first hand, in one's own soul, without any help from anyone other than God. If knowledge is based on the words of the "authorities" or the "experts," it is not realized knowledge, but imitative knowledge. It makes no difference if the authorities happen to be traditional prophets, like Moses, Jesus, and Muhammad, or modern-day prophets, like Darwin, Marx, and Einstein.

Some would respond that Muslims do not need to know things for themselves, because they can follow "consensus" (*ijmāʿ*), but this is true only in transmitted matters, not in intellectual matters. There is no such thing as *ijmāʿ* in the Islamic intellectual disciplines. Basic issues such as *tawḥīd* do not depend for their truth-value on the agreement of the ulama, as if the truth of a mathematical formula could be established by vote. Rather, the truth of the issues is self-evident to those who understand them.

One of the sure signs of the loss of intellectual knowledge is the strange phenomenon of Muslim thinkers apologizing for modern science by appealing to the "consensus" of the scientists. Even stranger is that they think they have taken an "intellectual" position. This shows that they have confused transmitted learning with intellectual learning. Modern science is indeed built on consensus, but this simply shows that it is fundamentally a transmitted science, not an intellectual science. Scientists do not verify and realize most of what they think they know. Rather, they accept it from their own authorities.

The truth of transmitted learning depends not on its self-evidence, but on the authority of its prophets and the reliability of its transmitters. It cannot be verified by individuals. Rather, it must be accepted on faith and trust, precisely because it is knowledge by imitation. For the Muslim intellectuals properly so called, the only possible way to know truth was to know it for oneself. When we do not know for ourselves, we have entered into the arena of transmitted beliefs. Modern science and learning is built on a vast structure of beliefs and presuppositions. The truth of its foundational beliefs is far from self-evident, and it certainly cannot be proven by the scientific method, given that the reliability of the method depends precisely on the presuppositions. The beliefs are part and parcel of a worldview, which is accepted on the basis of hearsay and consensus.

It can be argued that a modern scientist who makes a new discovery has "verified" and "realized" it for himself. The Muslim intellectual tradition would not have called this *taḥqīq*, however, because it does not extend deeply enough into the depths of the soul and spirit to recognize the real nature of things. Great scientific breakthroughs are achieved rather by what might be called "flashes of intuition," which pierce the limitations of consensual knowledge. On occasion this may be analogous to what the Sufi tradition calls "unveiling" (*kashf*), but the Sufi teachers always warn of the dangers of unveiling if it is not understood in light of the Qur'an and the Sunnah. However this may be, the flashes of insight necessary for scientific breakthroughs merely highlight the "prophetic" character of the great scientists. It says nothing about the gods from whom the revelations are received.

Whatever may be the exact nature of great scientific breakthroughs, the fact remains that the vast majority of scientists play the role of clerks, clerics, and workaday mullahs. In the very best cases, they are scientific *mujtahid*s, who apply scientific laws to new situations. The one thing a modern scientist or scholar can never be is a *muḥaqqiq*, a "realizer," unless he steps outside the context of his own discipline and allies himself with a living intellectual tradition.

In short, modern scientists – and, with far greater reason, the gullible public – accept scientific discoveries and "facts" on the basis of hearsay and consensus. They trust the promise that the discovery can be replicated by empirical research. They are usually unaware that modern theories are devices employed to interpret data for certain ends. They do not comprehend that the prestige of the theories derives not from their inherent truth, but from their usefulness for achieving certain specific ends and the degree in which they are accepted by the scientists, that is, the degree in which the scientific ulama reach consensus on the theory.

VERIFIABLE KNOWLEDGE

Having alluded to some of the profound differences between intellectual understanding and scientific findings, let me say something about the content of intellectual learning. What sort of knowledge can properly be verified and realized? What were Muslim intellectuals trying to know by themselves and for themselves, without following authority?

Note first that the purpose of the intellectual quest was not to gather information or what we call "facts." It was not to contribute to the progress of science, much less to build up a data base. Rather, its purpose was to refine human understanding. In other words, seekers of knowledge were trying to train their minds and polish their hearts so that they could understand everything that can properly be understood by the human mind, everything about which it is possible to have certain, sure, and verified knowledge. Each seeker of knowledge was trying to realize his knowledge for himself. He wanted to know his subject firsthand, with unmediated knowledge. If he took his knowledge from a teacher or a book instead of realizing its truth for himself, he was an imitator. Imitation can provide only transmitted knowledge.

Generally speaking, four major areas were considered the proper domains of realization: metaphysics, cosmology, spiritual psychology, and ethics.

Metaphysics is the study of the first and final reality that underlies all phenomena. The topic of discussion is God, though God is frequently called by impersonal names such as "Being," or "the Necessary," or "the Real."

Cosmology is the domain of the appearance and disappearance of the universe. Where does the universe come from, and where does it go? Naturally, it comes from the Real and goes back to the Real. But how exactly does it get here, and how exactly does it return? The intellectual tradition maintained that it was possible to verify the actual route of coming and going.

Spiritual psychology is the domain of the soul, the human self. What is a human being? Where do human beings come from, and where do they go? Why are people so different from each other?

How can people develop their potentialities? How can they become everything that they should and must become if they are to be fully human?

Finally, ethics is the domain of practical wisdom and interpersonal relations. How does one train one's soul to obey the dictates of intelligence, follow the guidelines of God, and carry out one's activities in harmony with the Real, the cosmos, and other human beings? What are the virtues that need to be achieved by a healthy and wholesome soul? How can these virtues become the soul's second nature?

It should be noted that the center of attention in all four domains was *nafs* – the self or soul. The human self is the key issue because it alone can come to know God and the cosmos. The way it does this is by developing and refining its own inner power, which is called "intellect" (*'aql*) or "heart" (*qalb*). If people are to develop and refine this power, they need to know what sort of self they are dealing with. You cannot know yourself by asking the experts to tell you who you are. You do not reach knowledge of yourself from outside, only from inside. Until you know yourself from within, your self-knowledge will be based on imitation, not realization.

All knowledge in the intellectual tradition was considered an aid in the process of coming to know oneself. The ancient maxim, "Know thyself" – often in the version attributed to the Prophet or 'Alī, "He who knows himself knows his Lord" – was taken seriously. The soul that is fully aware of itself is the soul that has perfected its potentiality as a knowing subject. In other words, through being fully conscious of its own reality, such a soul has become fully conscious of what God created it to be. The philosophers frequently called it *'aql bi'l-fi'l*, an actual intellect, or a fully actualized intellect. Such an intellect is nothing other than the soul that has perfected both its theoretical and its practical powers, both its vision and its virtue. Having become an actual intellect, the soul lives in harmony with God, the universe, and other human beings.

When the greatest masters of the tradition wrote about these four topics, they were writing about what they had realized, not simply what they had heard from someone else or reasoned out on the basis of someone else's theories or discoveries. They were critical of those who tried to grasp the issues merely on the basis of transmission, imitation, consensus, or argumentation. Intellectual questions demand intellectual answers, and the place to pose the questions and to understand the answers is within the self itself.

Among philosophers, Avicenna sets the tone of the quest when he describes the perfection of the soul in a passage found in two of his major philosophical statements, *al-Najāt* ("The Deliverance") and *al-Shifā'* ("The Healing"):

> The perfection specific to the rational soul is for her [the soul] to become an intellective world within which is represented the form of everything, the arrangement intelligible in everything, and the good that is effused upon everything, beginning from the Origin of everything and proceeding on to the unconditioned spiritual substances, then the spiritual [substances] connected in a certain way to bodies, then the high corporeal bodies along with their guises and potencies. Then [she continues on] like this until she fully achieves in herself the guise of all of existence. She turns into an intelligible world, parallel with all the existent world. She witnesses unconditioned comeliness, unconditioned good, and real, unconditioned beauty while being united with it, imprinted with its likeness and guise, strung upon its thread, and partaking of its substance.[2]

I will not try to unpack this passage here. Let me just note that the four major domains of philosophical inquiry are all present and that the focus is precisely on the soul that needs to be transformed into an actual intellect. Such a soul, coming to know itself through spiritual psychology, finds in itself metaphysical reality (the Origin), the whole cosmos in its various levels, and the realm of real and actualized ethics or virtue (the likeness of unconditioned good and beauty).

INTELLECT

The key to the Islamic *intellectual* tradition is precisely the intellect, which is nothing but the soul that has come to know and realize its full potential. Inasmuch as the soul possesses this potential, it is often called *fiṭra* or innate disposition. If we employ the language of the Qur'an, the *fiṭra* is the very self of Adam to whom God "taught all the names" (2:31). It is the primordial Adam present in every human being. At root, it is good and wise, because it inclines naturally toward *tawḥīd*, which stands at the heart of all wisdom and forms the basis for the acquisition of true knowledge of God, the universe, and the self.

The problem that people face with their *fiṭras* is that they are typically immersed in ignorance and forgetfulness. As long as the soul stays ignorant and forgetful of God, it cannot know its own *fiṭra* and cannot properly be called an "intellect." First, it must actualize its original, innate disposition and come to know all the names taught to Adam. Only then can it be called an "intellect" in the proper sense, that is, a fully actualized intellect.

To the extent that people fail to actualize their *fiṭra*, they remain ignorant of who they are and what the cosmos is. To the degree that they are able to actualize their *fiṭra*, they come to understand things in their principles, or in their roots and realities. In other words, they grasp things as they are related to God or as they are known to God. They do not remain staring at phenomena and appearances. Rather, they see with God-given insight into the real names of things. These names subsist eternally in the divine intelligence, which is the spirit that God blew into Adam after having molded his body from clay.

In short, the goal of the intellectual tradition was to help people come to know themselves so that they could achieve human perfection. To do so, one had to actualize both the theoretical intellect, which is the human self inasmuch as it knows all the realities and all the names, and the practical intellect, which is the human self inasmuch as it knows how to act correctly on the basis of the names taught by God.

From the perspective of this tradition, if we know things outside the divine context, we do not in fact know them. To the extent that we think we know them, we will be afflicted by compound ignorance. The more confident we are about the truth of our knowledge, the more difficult it will be to cure the disease. Moreover, it should be obvious that activity done on the basis of ignorance – not to mention compound ignorance – leads to ill consequences, not only for the individual, but also for society, the environment, and humanity.

BASIC FINDINGS

I repeat that, according to the masters of the intellectual tradition, you cannot gain intellectual understanding by listening to others or reading books. You have to find it in yourself. Nonetheless, it is useful to listen to what the great teachers have said in order to grasp the nature of their quest. When we do listen to them, we find that they agree on a large number of points, though they tend to use a diversity of expressions. Mentioning a few of these points can help us understand what exactly premodern Muslim intellectuals were trying to verify and realize. Let me list ten of them:

1. *Tawḥīd.* All reality is unified in its principle. Everything in the universe comes from God and returns to God, and

everything is utterly and absolutely dependent upon God here and now, always and forever, in every time and in every place.

2. The eternal light of God is a permanent presence in the created order. The reverberation of this light in human experience is called "intellect" or "spirit" or "heart." All things are known to this light, because it is the conscious and aware pattern in terms of which both the universe and human beings came into existence.
3. The cosmos is a grand hierarchy in which every level of reality is present simultaneously, without regard to temporal succession. This hierarchy is ordered in an intelligent way, according to the wisdom of God, and it begins and ends in the light of God.
4. This hierarchical cosmos is divided into two basic worlds, one visible and one invisible. The invisible world is the domain of spirit, light, intelligence, and awareness. The visible world is the domain of body, darkness, ignorance, and unconsciousness. The invisible realm is closer to God and more real than the visible world. The visible, physical realm is the most amorphous, least intelligible, and least substantial of all real domains. Given its relative unreality and its subservient status, the physical realm has no control over the spiritual realm, just as created things have no control over God.
5. Human beings are unique in the cosmos. God made them in his own image and taught them all the names. As a result, everything found in the external universe is also found, in essence and reality, in the primordial human selfhood known as *fiṭra.*
6. The cosmos is animated by two, simultaneous movements. First is the centrifugal movement away from the Source, second the centripetal movement toward the Source. These are what the philosophical tradition commonly called *al-mabda'*

wa'l-ma'ād, "the Origin and the Return." The Sufis often used the expression *al-qaws al-nuzūlī* and *al-qaws al-ṣu'ūdī*, "the descending arc and the ascending arc." The issues discussed are cosmogenesis and eschatology. When addressing the fact that all things naturally and necessarily go back to their Origin, the tradition also discusses the uniquely human privilege of voluntarily returning to God. Freely choosing to return is precisely the *raison d'être* of realization, and realization is another name for the voluntary return. Or, in Sufi language, the voluntary return provides the means to "die before you die."

7. Despite these two movements – centrifugal and centripetal, descending and ascending – intelligence per se never leaves its own invisible and transcendent realm. In its deepest nature, the human self is indistinguishable from intelligence, so it remains indefinable and non-specific. Every specific thing and every specific viewpoint tells the self what it is not. The self knows that it is not limited by the objects of its knowledge or by the finiteness of things, nor by the limitations of this standpoint or that science; it also knows that it has the potential to perceive and comprehend all definitions and all limitations. Hence it knows – if it is self-aware – that it has no inherent limitations. It knows that it is free, not of this or that, but of all things, of everything other than the Real.
8. The final goal of religion, and, indeed, of all human endeavor, is to awaken the intellect. Awareness of whatever sort is nothing but a glimmer of intellect, and there are infinite degrees of awakening. People are diverse in their aptitudes for finding the divine light within themselves. The teachings of prophets, sages, and avataras are addressed to all people and are meant to guide everyone to the light; following these teachings properly and sincerely will ensure that people find the light in a

congenial manner after death. The intellectual tradition is designed to guide those who have the capacity to develop their self-awareness through realization here and now, without waiting for the promises of the afterworld.

9. Our individual selves are identical with our awareness of things. We are what we know. The fullness of our original, innate disposition – our *fiṭra* – is found in the fullness of understanding. The more we understand, the more human we are. The more forgetful and heedless we are, the less human we are. To the degree that we imitate others in *intellectual* knowledge, we will fail to actualize our *fiṭra* and move further from human perfection.
10. The theoretical and practical sides of the intellect need to be developed in harmony. The role of the theoretical side is to know things as they truly are, and the role of the practical side is to discern proper activity and beautiful behavior.

A VISITOR FROM THE PAST

Having taken a quick look at the intellectual tradition, let me perform a thought experiment. It is commonly imagined that if our ancestors could be brought from the past in a time machine, they would be amazed and dumbfounded by the feats of modern science and civilization. But how would a Muslim intellectual of the past react to the modern world, and in particular to its intellectual ambience? What would an al-Fārābī, or an Avicenna, or a Mullā Ṣadrā think of contemporary science and scholarship?

For the purpose of this experiment, I will borrow the name of our time-traveler from the famous philosophical novel of Ibn Ṭufayl, *Ḥayy ibn Yaqẓān*, "Alive, son of Awake." The name refers to the soul that has been reborn by actualizing the intellect. I will simply call him Ibn Yaqẓān.

No doubt Ibn Yaqẓān would be astonished by the ready availability of an enormous amount of information. However, he would be much more astonished by the fact that people have no idea that all this information is irrelevant to the goal of human life. He would see that people's understanding of their true situation has decreased roughly in proportion to the amount of information they have gathered. The more "facts" they know, the less they grasp the significance of the facts and the nature of their own selves and the world around them.

Ibn Yaqẓān would be appalled at the loss of any sense of what knowledge is *for*. People think that they should gain knowledge to control their social and natural environments and to make their physical lives more comfortable. In Ibn Yaqẓān's view, the "quest for knowledge" that the Prophet made incumbent upon all believers is not, however, a quest for information or a "better life." Rather, it is a quest to understand the Qur'an and the Hadith, and then, on the basis of that understanding, it is a search for self-knowledge, self-awareness, and the recognition of God's signs in the universe and the soul. It is a quest for wisdom and mastery of self, not for control and manipulation of the world and society.

Ibn Yaqẓān would certainly be struck by the misuse of words like "scientist" and "intellectual." He would immediately see that people use the word "scientist" to designate possessors of a knowledge that is deemed uniquely true and reliable. He would see, however, that "scientific" knowledge is simply a means for understanding appearances so that they can be manipulated to achieve the desires of human egos. To him, it would seem that what people call "science" is strikingly similar to what in his times was called "sorcery." Certainly, the goal is exactly the same: to manipulate God's creation by recourse to means that escape ordinary human abilities for the sake of short-sighted if not demonic goals.

As for the word "intellectual," he would think that an intellectual is someone who knows God, the world, and the human soul on the basis of realization, not imitation. An intellectual is someone who claims to know only what he has realized for himself, and otherwise quotes the authorities or admits his ignorance. Ibn Yaqẓān would see, however, that modern scientists, intellectuals, and scholars have acquired all their knowledge by imitation, not realization. They take what they call "facts" from others, without verifying their truth, and then they proceed to build their own theories and practices on the basis of these borrowed facts, producing an endless proliferation of new facts that go back to no firm foundation. Experts in the modern scientific and critical disciplines do not know things as they are, but rather in terms of the consensus of their colleagues, mathematical constructs, theoretical fantasies, and ideological presuppositions.

Ibn Yaqẓān would think that the modern learned classes imagine that they know all sorts of things, but in fact they know nothing. Verified and realized knowledge carries with it the self-evidence of certainty, but people have no certainty about anything. Since all their information and learning is of the transmitted variety, they do not know for themselves and in themselves.

Ibn Yaqẓān would be amazed at the blatant polytheism that drives mental and social endeavor. He would see that the modern world asserts a great multiplicity of gods with respectable, scientific-sounding names like development and progress. Instead of a worldview of *tawḥīd*, he would see a worldview of *takthīr*. He would quickly understand that the diverse technical, scientific, social, and political solutions that are offered to bring peace and harmony to the world simply intensify the reign of *takthīr*.

Ibn Yaqẓān would be astonished that even scholars and scientists who consider themselves religious are convinced that the only way truly to know something is to begin with the many, not the

One. In his intellectual tradition, all thinking began with an understanding of the Primal Unity that lies beneath and beyond surface multiplicity and gives meaning to all things – from stars and celestial phenomena to minerals and plants, from prophetic teachings to logic and mathematics. But he would see religious people claiming that modern science does not contradict the Primal Unity because it is simply a "method," or a way to understand mechanisms and workings. He would wonder at a blatant polytheism that thinks that there can be any real understanding of the many apart from the One. How can we deal with methods and mechanisms without reference to the Creator of the mind that devises the methods and mechanisms and without reference to the goals and aspirations of the devising mind? All this is to set up a series of independent realities. And to set up realities, objects, and methods without demonstrating explicitly how these are subservient to the laws of the One is precisely *takthīr.*

Along with a multiplicity of gods called by abstract, respectable names, Ibn Yaqẓān would see ranks upon ranks of priests serving the gods and encouraging their followers to immerse themselves in dispersion and confusion. He would see that each priesthood jealously guards its esoteric knowledge from the common people. He would also notice, however, that the common people – who consider themselves among the enlightened few in history – no longer believe in priests. Hence the priests call themselves doctors, surgeons, physicists, biologists, engineers, sociologists, political scientists, programmers, lawyers, professors, and experts. Ibn Yaqẓān would be surprised that so many people think that these priests have a sacred, transmitted knowledge that is worthy of imitation and blind obedience.

Ibn Yaqẓān would be coming from a religious tradition that has a dim view of priests in the first place. He would not be surprised to see that each contingent of priests contends with the others for a greater share of wealth, prestige, and social control. He would perhaps be impressed by the enormous sanctuaries that they build for themselves in the names of their gods, the great cathedrals of Medicine, Technology, and Scholarship. However, he would be horrified by the ugliness of the buildings and the unspeakable rituals that some of the priests force upon their followers, such as the last rites reserved for believers in Medicine.

To make a long story short, Ibn Yaqẓān would be appalled not only by the misguided beliefs of the common people, but also by the sophisticated *takthīr* of the learned classes. In both cases, he would see that people have lost any sense of what is truly real. He would be shocked by the way people immerse themselves in meaningless hopes and illusory endeavors. He would be dismayed by the willful blindness toward the permanent, everlasting, omnipresent reality that is the intelligent and intelligible light of God. He would be aghast at the loss of any sense of the hierarchical structure of the cosmos and the soul, at the flattening of the world that makes material appearance seem to be the only reality. He would be astonished that people have surrendered their freedom to the esoteric knowledge of priests. He would be amazed that a class known as "intellectuals" thinks that *tawḥīd* and all that was considered worthy of aspiration in past times were misguided delusions, self-serving fantasies, rationales for social injustice, and epiphenomena of psychological contingencies.

As for Muslims living in the modern world, he would be dumbfounded that most of them accept the gods and priests just like the non-Muslims. What would perhaps sadden him most, however, is that Muslim parents have lost any sense of how to guide their children on the path of *tawḥīd*. They have come to believe that religion means ignorance and superstition, and that studying the Islamic heritage is a total waste of time, since it has been replaced by scientific knowledge. They refuse to allow their children to study religion except when all other avenues of advancement are barred. Medicine, science, engineering, and business administration are the professions of choice, and – in North America at least – law, since lawyers make a lot of money too. So, instead of encouraging their children to search for knowledge of God and his guidance, they insist that they join one of the priesthoods. The learning that their children gain is still of the transmitted variety, but joining the priesthood of doctors is much more respectable – not to mention lucrative – than becoming a Muslim cleric.

After taking a quick look around, Ibn Yaqẓān would no doubt be anxious to return to a world that has preserved some sense of proportion.

3

The Rehabilitation of Thought

Few authors have left as deep an impression on modern-day Muslim thinking in the Indian subcontinent as Allama Iqbal. Given his laudable efforts to reformulate the basic theoretical teachings of Islam in a manner that would be appropriate for modern times, especially his *Reconstruction of Religious Thought in Islam*, I would like to take the occasion of this lecture in a series named after him to reflect on thirty-five years of studying Islamic thought.[3] The questions I asked myself in preparing the talk went something like this: is there anything about traditional Islamic thought that makes it more than an historical curiosity? Is it relevant to the very real and concrete problems that all human beings, not just Muslims, face at the beginning of the twenty-first century? Should Muslims continue the common practice, acquired in the nineteenth and twentieth centuries, of ignoring their own intellectual heritage in attempting to reformulate Islamic teachings?

My general answer to these questions is that Islamic thought is indeed far more than an historical curiosity. It is a valuable repository of profound teachings about the human predicament. Not only is it relevant to contemporary concerns, it is far more relevant than any of the sciences, technologies, and ideologies that occupy the minds of most contemporary thinkers, Muslim or otherwise. In fact, traditional Islamic thought is so relevant to Muslim attempts to deal with contemporary issues that, if it is not recovered and rehabilitated, authentic Islamic thinking will cease to exist. In other words, there will be no escape from what dominates most contemporary Islamic thought already, which is warmed-over ideology disguised by a veneer of Islamic rhetoric.

If genuine Islamic thought ceases to exist, the religion of Islam will lose touch with its living roots and no longer function as an alternative to modernity. One might think that this would be a good thing – is this not precisely what the reconstruction of Islamic thought is all about? The problem here is that modernity is propelled by a certain type of false thinking that is intensely antithetical to the three principles of Islamic faith – *tawḥīd*, prophecy, and the Return to God. The antidote to false thinking is not blind faith in new forms of transmitted knowledge, but rather true thinking. Any sort of true thinking must be anchored in the nature of reality itself, which is expressed Islamically in the three principles.

To think in Islamic terms one needs to reconnect one's thought to the transcendent truths from which Islam draws sustenance. This needs to be done not only by having recourse to the guidelines set down in the Qur'an and the Hadith, but also by seeking help from the great Muslim intellectuals of the past, those who employed the Qur'an and the Hadith to clarify the proper role of thought in human affairs.

THOUGHT

To explain what I mean by the proper role of thought, I need to recall the primary position given to thought throughout Islamic history. By "thought" I mean the human ability to be aware of things and to articulate this awareness in concepts and language. For those familiar with the Islamic worldview, it is not too difficult to see that thought has always been considered the single most important component of human life, and that it must be attended to before all else.

The primacy of thought is made explicit in the first half of the Shahadah, the testimony of faith: "There is no god but God." This is the one truth upon which all of Islam depends. The *tawḥīd* that is expressed here is not contingent upon the facts and events of the world. It is essentially a thought, a logical and coherent statement about the nature of reality. In the Qur'anic view of things, *tawḥīd* guides the thinking of all human beings inasmuch as they are true to their innate disposition (*fiṭra*). Every messenger from God came with *tawḥīd* in order to remind his own people of their humanity. In this way of looking at things, true thought is far more real than the bodily realm, which is nothing but the apparition of thought. This is not to say that the external world has no objective reality, far from it. It is to say that the universe is born from the consciousness, awareness, and thought of the divine and spiritual realms.

It should be obvious that by real thought I do not mean the superficial activities of the mind, such as reason, reflective thinking, ideation, cogitation, and logical argumentation. Rather, I mean the very root of human existence, which is consciousness, awareness, and understanding. The Islamic philosophical tradition usually referred to this root as *'aql*, intelligence. Thought in this sense is a spiritual reality that has being and life by definition. In contrast, the bodily realm is essentially dead and evanescent, despite the momentary appearance of life within it. Intelligence is aware, but things and objects are unaware. Intelligence is active, but things are passive. Intelligence is living, self-conscious, and dynamic, but things are empty of these qualities in themselves. In its utmost purity, intelligence is simply the shining light of the living God, a light that bestows existence, life, and consciousness on the universe. It is the creative command whereby God brought the universe into being, the spirit that God blew into Adam after having molded his clay, and the divine speech that conveys to Adam the names of all things.

Islamic forms of thinking take it for granted that God is the source of all reality. The universe and all things within it appear from God in stages, just as light appears from the sun by degrees. The spiritual world, which the Qur'an calls the Unseen (*al-ghayb*), is the realm of life, awareness, and intelligence. The bodily world, which the Qur'an calls the Visible (*al-shahāda*), is the realm of death, unawareness, and unintelligence. The closer a creature is situated to God, the more immersed it is in the light of intelligence, consciousness, and thought. Angels and spirits, who inhabit the Unseen, are vastly more intense in luminosity and intelligence than most inhabitants of the visible realm.

In this way of looking at things, human beings, who were placed on the earth to be God's vicegerents (*khalīfa*), are nothing but thought. Their awareness and consciousness determine their reality. Their thoughts mold their nature and shape their destiny. The great Persian poet Rūmī, a true master of the intellectual tradition, reminds us of thought's primacy in his verses,

> Brother, you are this very thought –
> the rest of you is bones and fiber.
> If roses are your thought, you are a rose garden,
> if thorns, you are fuel for the furnace.
> If rosewater, you will be sprinkled on the neck,
> if urine, you will be dumped in the pit.[4]

It is human nature to understand that we are essentially thought and awareness, but we forget it constantly. We are too preoccupied with our daily activities to stop and think. We are too busy to remember God and apply the principle of *tawḥīd* to life, a principle that guides all true thought back to the One Origin of thinking. Without the constant reorientation of thought by the remembrance of the One, people can only forget their innate human disposition.

If thought determines our present situation and our final outcome, what about the content of thought? Toward what end should thought be directed? The position of the Islamic tradition has always been that thought must be focused on what is real, and nothing is truly real but God, the Real (*al-ḥaqq*). The whole activity of thought must be ordered and arranged so that it begins and ends with the supreme reality. Moreover, moment by moment, thought must be sustained by awareness of the Real. Forgetting God, one needs to recall, was Adam's sin. In his case, the sin was quickly forgiven, because he immediately remembered. But most people do not remember, especially in modern times, and the consequences have been disastrous. As the Qur'an puts, "They forget God, so God forgets them" (9:67). Being forgotten by God is to be cut off from the awareness of Reality and to fall into illusion and unreality.

True thought, then, accords with the divine spirit that lies at the core of human awareness. It is to understand things as they are. Things can only be understood *as they are* if one is aware of them in relation to the Creator who sustains them moment by moment. True thought is to see things in relation to God. This is precisely the meaning of *tawḥīd*.

Rūmī tells us repeatedly about the proper object of thought, and he often reminds us that true thought is living intelligence, or another kind of vision. Take these verses:

> To be human is to see, the rest is only skin.
>   To see is to see your beloved.
> If your beloved is not seen, better to be blind.
>   If your beloved is not everlasting, better not to have one.[5]

Rūmī is saying that human beings are governed totally by their awareness of goals and desires. Any thought, any vision, any understanding that is not informed and guided by the awareness of God's overwhelming and controlling reality loses sight of the nature of things and forgets the purpose of human life.

THE INTELLECTUAL TRADITION

In speaking of traditional Islamic thought I mean intellectual, not transmitted, learning. As noted already, four main areas of inquiry dominated the concerns of the Muslim intellectuals: metaphysics, cosmology, spiritual psychology, and ethics. As for the various branches of intellectual learning that resembled what we nowadays call "science," they focused on secondary issues pertaining to cosmology. Most Muslim intellectuals were not interested in such issues per se, but only inasmuch as they could throw light on the primary topics.

The basic characteristic of Islamic intellectuality was its unitary vision. The various sciences were not understood as separate and independent realms of inquiry, but as complementary domains. The more one investigated the external world – the domain of cosmology – the more one gained insight into the internal world, the domain of spiritual psychology. The interrelationship among the fields of intellectual inquiry is especially obvious in these two realms.

On the philosophical side of the intellectual tradition, the importance of the interrelationship between cosmos and soul is already apparent in the expression *al-mabda' wa'l-ma'ād*, "The Origin and the Return," which was prominent enough to be the title of books by both Avicenna and Mullā Ṣadrā, arguably the two greatest Muslim philosophers. As Islamic philosophy developed, the return of the soul to God, *al-ma'ād*, became more and more the focus of attention. Although Western scholars usually translate this term as "eschatology," the philosophers who discussed it were not primarily concerned with death, afterlife, and the resurrection. Rather, they wanted to understand and explicate the nature of the ongoing and ever-present human ascent toward God.

Moreover, even though metaphysics and cosmology center on God and the cosmos, both were studied with the aim of understanding the true nature of the human soul. The simple reason for this is that we cannot understand ourselves without understanding God and the universe. Only in terms of a true comprehension of the nature of things can people orient themselves in relation to their own ultimate concerns. Only on the basis of a correct orientation can they set out to achieve the goal of human life, which is to be completely human.

In short, the purpose of intellectual studies was to prepare the ground for achieving human perfection. Perfection can only be reached by "returning" to God, that is, by bringing oneself back into harmony with the true nature of things. Both philosophers and Sufis were striving to become what it is possible for human beings to become. To use the expression that was made famous by Ibn 'Arabī, the goal of human life was the achievement of the status of *insān kāmil*, "a perfect human being."

TAQLĪD AND *TAḤQĪQ*

In his attempts to reconstruct Islamic thought, Allama Iqbal was much concerned with overcoming *taqlīd* or imitation and reviving *ijtihād*, the independent judgment that allows a person to make sound legal decisions on the basis of the Qur'an and the Hadith. But, as he well knew, the word *taqlīd* has two opposites in the Islamic sciences. If we are discussing jurisprudence and the Shariah, its opposite is *ijtihād*, and Islamic law holds that Muslim believers have the duty either to follow someone else's *ijtihād*, or to be *mujtahid*s themselves. In the intellectual sciences *taqlīd*'s opposite is *taḥqīq*, verification or realization.

Taḥqīq derives from the same root as *ḥaqq*, which means truth, reality, appropriateness, rightness, responsibility, and duty. *Taḥqīq* means not only to understand the truth, rightness, and appropriateness of things, but also to respond to them correctly by putting into practice the demands that they make upon the soul. By its nature, understanding of any kind is intensely personal. One can understand the *ḥaqq* of things only for oneself and in oneself. A *muḥaqqiq* is someone who knows without the intermediary of transmission and acts appropriately. He fulfills his responsibility toward God, creation, and society on the basis of a verified and realized knowledge, not on the basis of imitating the opinions and activities of others.

When great Muslims of the past, such as Rūmī or al-Ghazālī, criticized *taqlīd*, they were not criticizing imitation of the ulama in matters of the Shariah. Rather, they were attacking *taqlīd* in questions of understanding. You cannot understand God or your own self by quoting the opinions of others, not even if the others are the Qur'an and the Prophet. The only way to understand things is to find out for yourself, even though you need the help of those who already know. The goal was to allow people to think properly, not to follow someone else's thinking. On the basis of proper thought, people can reach a correct understanding of the objects that pertain strictly to intelligence. The first and most important of these objects is *tawḥīd*, the one truth that underlies every other truth.

The real disaster that looms over the Islamic tradition has little to do with *ijtihād* and everything to do with *taḥqīq*. A society without living *mujtahid*s can continue to function more or less adequately on the basis of imitating the scholars of the past. A society without living *muḥaqqiq*s, however, has surrendered the ground of intelligence. It cannot hope to remain true to its principles, because it cannot *understand* its principles. What I am saying is that *tawḥīd* can only be understood through realization, not imitation and certainly not through *ijtihād*. Once Muslims lose sight of their own tradition of understanding, they have lost the ability to see with the eye of *tawḥīd*.

To lose the ability to see with the eye of *tawḥīd* means to fall into seeing with the eye of *shirk*, or associating other gods with God. If the Qur'an considers unrepented *shirk* the one unforgivable sin, this is no doubt because it entails an utter distortion of human understanding, a corruption of the human *fiṭra*, and an obscuration of the intelligence that is innate to every human being.

Given that *tawḥīd* is the primary duty of every Muslim, and given that *tawḥīd* can be defined negatively as "the avoidance of *shirk*," it follows that avoiding *shirk* is the primary duty of every Muslim. And, just as *tawḥīd* is the first principle of right thinking, so also *shirk* is the first principle of wrong thinking. In other words, *shirk* is an intellectual issue, just as *tawḥīd* is an intellectual issue. Any form of thinking that is not rooted in *tawḥīd* necessarily participates in *shirk*.

PREMODERN SCIENCE

By mentioning the "rehabilitation" of Islamic thought, I mean to suggest that the Islamic intellectual tradition is suffering from a grave illness. Although a good deal of thinking goes on among contemporary Muslims, little of it has roots in the Islamic intellectual tradition. It frequently calls upon the Qur'an and the Hadith as witness, but it is based on habits of mind that were developed in the West during the modern period. These habits of mind, if judged by the principles of Islamic thinking, are misguided and wrong-headed. In other words, they are rooted in *shirk*, not *tawḥīd*.

If we accept that traditional Islamic thought is gravely ill, it will be obvious that recovery demands intensive care. Among other things, it will involve a thorough re-evaluation of the nature of intellectual health. It will necessitate careful scrutiny of the great texts of Islamic philosophy and theoretical Sufism and a serious attempt to understand Islamic principles by way of realization, not imitation.

Before rehabilitation can begin, the illness must be correctly diagnosed. The diagnosis of an intellectual illness depends upon recognizing error for what it is. The problem here is that the illness is omnipresent, not only in the Islamic world, but also elsewhere. It is so much a part of the way that most people think that they imagine it to be natural and normal. Like someone suffering from a debilitating disease since childhood, people have lost any sense of what health might involve. This disease is co-extensive with the worldview that informs modern thought.

It is very difficult to characterize the modern worldview with a single label. One word that has often been suggested is "scientism," the belief that the scientific method and scientific findings are the sole criterion for truth. Like most belief-systems, scientism has become second nature to its believers. It is a basic characteristic of the modern worldview and the contemporary *Zeitgeist*. People see the world and their own psyches in terms of what they have learned in schools, universities, and television documentaries. It is simply assumed that the universe described by science is the real universe. If religious teachings are taken seriously, they are understood as pertaining merely to ritual and morality, not to the "real world," since only science provides reliable knowledge of the universe.

One of the many implications of the scientistic worldview is the common belief that the cosmology and natural sciences developed in the Islamic intellectual tradition were early stages of what we nowadays call science, and that most of these early findings have now been proven false. But a basic fallacy informs this view of premodern science: the assumption that its aims and goals were the same as those of contemporary science. If this were true, then indeed the premodern ideas would be incorrect. However, the fact is that the Muslim scientists, all of whom were trained in the intellectual tradition, were busy with a task that is far different from that which occupies modern scientists. In order to understand the Islamic intellectual tradition, it might be better to avoid altogether the use of the word "science" to designate what they were doing, given that this word has been pre-empted by the empirical methodologies that characterize the modern period. Instead, we need to recover a term that represents the real goal of Muslim intellectuals.

One possible name for both the methodology and the goal of this tradition, a name that was in fact commonly employed, is *ḥikma* or "wisdom." This word has the advantage of not implying a scientific and empirical approach to things, and it also has the advantage of being a divine attribute. In English, it makes perfect sense to say that God is Wise, but not that he is Scientist. The English word "wisdom" and the Arabic word *ḥikma* have preserved enough of their ancient meaning to imply both right thought and right activity, both intellectual perfection and moral perfection.

In contrast, modern scientists long ago abandoned any claim that science can help people find the road to right activity, not to speak of moral perfection. The role of science is simply to provide more power over God's creation. Science does not and cannot address the issue of understanding the true nature of the universe, because the true nature of the universe cannot be understood without reference to the transcendent, intelligent, unseen principles that govern the universe. Nor can science address the issue of how we are to find the wisdom to employ correctly the power that we gain over creation. That is the job, scientists will tell us, of theologians, moralists, and politicians.

Another name that fairly describes the goal of Islamic thought is *taḥqīq*. The focus of Muslim intellectuals was not on the practical affairs of this world, but on the full realization of human intelligence. This demanded not only discovering the *ḥaqq* of things, their truth and reality, but also acting in accordance with that *ḥaqq*. This could only be determined by reference to *al-ḥaqq*, the Real, the absolute reality that is God. *Taḥqīq* demands both right thought and right activity, both intellectual perfection and moral perfection.

The Islamic quest for wisdom was always a quest to achieve unity with the divine light or the divine spirit. By the nature of the quest, Muslim intellectuals knew from the outset that everything had come from the One and will return to the One. Their quest was not to "believe" that God is one, because they already knew that God is one. The unity of Ultimate Reality was too self-evident to be doubted. The quest was to understand the implications of unity thoroughly and completely.

In brief, the purpose of searching for wisdom was what we can call "the *taḥqīq* of *tawḥīd*," and it had two complementary dimensions: it meant first to verify and realize the truth of *tawḥīd* for oneself, and second to put that truth into practice in all thought and activity. The goal, in other words, was spiritual transformation. This was understood to involve a total conformity with the divine attributes (*ṣifāt*) and character traits (*akhlāq*). It was often called *ta'alluh*, "deiformity," or *takhalluq bi akhlāq Allāh*, "assuming as one's own the character traits of God."

Tawḥīd was considered both the seed and the fruit of human possibility. It was the seed that was planted in human awareness, and it was the fruit of the soul's tree – perfect understanding and perfect activity. In such a view of things, it was impossible to separate the realms of learning into independent domains. The *taḥqīq* of *tawḥīd* was a holistic enterprise that yielded a unified vision. This vision demanded the unity of the human subject with the cosmic object, that is, the conformity of the full human soul with the cosmos in all its grandeur. Soul and cosmos were seen as complementary manifestations of the One, Single Principle. When God created Adam in his own image, he also created the universe in his own image. Perfect understanding means the ability to see all things in their proper places, as divine images and in their relationship with their Source.

The basic position of the tradition was always that understanding the knowing self, the subject that takes the cosmos as its object, was essential to the quest. It was impossible to ignore the self or to pretend that it was anything other than an integral part of a greater whole. It is here in particular that the Western tradition diverged from the Islamic. Any careful investigation of the great thinkers of the Enlightenment, the fathers of modern thought, shows that they completely ignored the complementarity of soul and cosmos. Bryan Appleyard does a brilliant job of analyzing this phenomenon in his *Understanding the Present.* For example, he writes,

> Protestantism and the Renaissance had effectively prepared the way: the first by insisting on the moral centrality of the individual and the second by its celebration of heroic humanism. The price was the expulsion of the self from the world. For science made exiles of us all. It took our souls out of our bodies.
>
> The tendency is evident in the primary philosophers of the Enlightenment. Descartes provided a philosophical correlative of Protestant internalization. ... Kant removed the real world beyond the possibility of ordinary human knowledge. Both placed the world that was the object of scientific investigation beyond the realm of the self. The key paradox of the modern was established: science was everything we could logically know of the world, but it

> could not include ourselves. ... The more we knew, the less we appeared to have a role. The world worked without us.[6]

THE REIGN OF *TAKTHĪR*

I said earlier that modernity is governed by a certain type of false thinking, and I suggested that one name for that thinking is "scientism," which is false because it makes unwarranted claims. But there is a much deeper reason why scientism is essentially false, and that is because science, by its very presuppositions, negates *tawḥīd* and affirms *takthīr*.

By no means do I mean to say that *takthīr* is inherently false. Rather, it is short-sighted and incomplete. It misses the important points, because it denies implicitly, if not explicitly, the ultimacy of the One Reality that stands beyond all other realities. Once we understand things in terms of *tawḥīd*, we can understand the origin and destiny of the cosmos and the soul, and we can also grasp the present status of the world in which we live. *Tawḥīd* answers the ultimate questions and allows people to orient themselves in terms of real beginnings and real ends.

If *takthīr* is to have any legitimacy, it must be oriented and governed by *tawḥīd*. *Takthīr* without *tawḥīd* can at best analyze, differentiate, divide, and classify, but it cannot provide a unifying vision. Any perspective based on *takthīr* denies implicitly that existence has a purpose. It rejects the idea that human aspirations to achieve moral and ethical betterment and to become intellectually and spiritually perfect have any grounding in objective reality. Consequently, this perspective means that the more *takthīr* is intensified, the less we as human beings will appear to have any role at all to play in the cosmos.

The Muslim cosmologists paid a good deal of attention to *takthīr*, but for them it was a divine attribute. It was God's activity of bringing the universe into existence. When they investigated the Origin of all things, they were attempting to understand the nature of *takthīr*. In effect, they saw God as *al-mukaththir*, "He who produces the many." In contrast, when they discussed psychology and the return of the soul to God, the primary issue was how the soul could be a *muwaḥḥid*, "someone who affirms the One, who establishes Unity." How can we, beings who dwell in multiplicity, unify our vision and activity and return happily and freely to God?

In the intellectual tradition, we can understand *takthīr* as the divine principle that makes multiplicity appear from the One. *Tawḥīd* can then be understood as the complement of *takthīr*. It designates the divine and human principle that reintegrates the many into the One. The philosopher Afḍal al-Dīn Kāshānī, for example, tells us that the Universal Intellect is God's vicegerent in the Origin, which is to say that the cosmos in all its multiplicity appears from unity on the basis of the radiance of the divine omniscience. In contrast, human beings are God's vicegerent in the Return, which is to say that the human role in the cosmos is to take multiplicity back to unity.[7]

In brief, the intellectual tradition recognizes both *takthīr* and *tawḥīd*, but *takthīr* is kept subordinate to *tawḥīd*, which is to say that the many is seen as forever governed by the One. The world and all things within it stay in the hands of the Real and can never leave. The proper role of *takthīr* can only be understood in terms of *tawḥīd*. Once we see that God created human beings to act as his vicegerents and unify the whole of creation through their spiritual and moral perfection, then we can understand why God brought multiplicity into existence in the first place. Real understanding and real knowledge depend on grasping the ultimate end of human existence, which corresponds with the ultimate end of creation itself. Moreover, human completion and perfection depend on acting in conformity with real knowledge.

The Islamic worldview might be characterized as *takthīr* in the service of *tawḥīd*. In contrast, the scientific worldview can be characterized as *takthīr* without *tawḥīd*. This can be seen clearly in the fruit of modern learning. Take, for example, the ever more specialized nature of the scientific, social, and humanistic disciplines; the disintegration of any coherent vision of human nature in the modern university; the unintelligibility of the individual sciences to any but the experts; and the total incomprehensibility of the edifice of science and learning as a whole. When *takthīr* rules over human thought, the result can only be analysis, differentiation, distinction, disunity, disharmony, disequilibrium, and dissolution. Given that modern science and learning are rooted in the world's multiplicity, not in God's unity, their fruit is division and dispersion without end, not unification and harmony.

One of Iqbal's great insights, which he did not follow up as he might have, was his understanding that modern science yields disunity and dissonance by definition. He wrote,

> We must not forget that what is called science is ... a mass of sectional views of Reality. ... [T]he various natural sciences are like so many vultures falling on the dead body of Nature, and each running away with a piece of its flesh. Nature as the subject of science is a highly artificial affair, and this artificiality is the result of that selective process to which science must subject her in the interests of precision.[8]

Modern science wants "precision" in order to separate things out from their overall context. Only after a "highly artificial" view of reality has been manufactured can we ignore the objectivity of moral and ethical principles and justify the view that human beings have the right to control God's creation as they see fit, without the guidance of wisdom.

Perhaps the power of *takthīr* becomes most obvious in the realm of ethics and morality. For the Islamic intellectual perspective, adherence to right activity and actualization of "praiseworthy character traits" (*akhlāq ḥamīda*) are demanded by the objective nature of things. After all, the world is actually and truly a display of the divine attributes, and the human soul is actually and in fact made in God's image. Any human soul that does not actualize the divine character traits – such as wisdom, justice, mercy, compassion, love, and forgiveness – has failed in the task of achieving human status.

A methodology that yields an unbridgeable gulf between truth and ethics is ignorance, not knowledge. Such an approach ignores the *ḥaqq* of things – both their true nature and the moral demands that they make upon us. Under the reign of *takthīr*, intelligence and virtue are torn from their roots in the real world. The net result can only be the dispersal of human excellence in a vast range of unrelated endeavors, with no connections to be made between knowing and being, science and ethics. The raw power that is accumulated through acquiring instrumental and manipulative knowledge results in the loss of human goodness.

THE GOAL OF THOUGHT

I said that there is a fundamental difference between the Islamic intellectual tradition and modern learning. One way to understand this is to see that Muslim intellectuals were striving to achieve a unitary and unified vision of all things by actualizing the transpersonal intellect, the divine spirit latent in the human soul. In contrast, modern scientists want to achieve an ever more exact and precise understanding of things, one that allows for increased control over the environment, the human body, and society. To the extent that this control is achieved, however, it is given over to the ignorant and forgetful selfhood – what was called "caprice" (*hawā'*) or "appetite" (*shahwa*) in the texts. It is not put into the hands of the fully actualized intelligence of God's vicegerent on earth. This is especially obvious in the various forms of government that have appeared in the modern world, all of which take advantage of scientific, technological, and bureaucratic power to instill docility into their subjects.

Another characteristic of the intellectual tradition that places it in stark contrast with modern learning is the intensely personal nature of the quest. *Taḥqīq* aims at the discovery of the *ḥaqq* within the seeker's own intelligence. That intelligence was understood, and, indeed, experienced, as the supra-individual, transpersonal, universal breath of awareness. Every seeker of wisdom had to learn metaphysics and cosmology for himself or herself. Each had to follow the path of self-discovery as a personal calling. In other words, aspiring philosophers had to relearn the nature of the cosmos for themselves, not depend on what was written in the authoritative texts. From a modern perspective, it looks like they were trying to "reinvent the wheel." Implicit in the metaphor is the technological application of knowledge that is a primary motivation for scientific research and was in no way part of the quest for wisdom. Actualizing wisdom can only be achieved in realization, which is awakened intelligence and ethical activity.

It is a common misinterpretation of Islamic intellectual history to say that Muslim scholars made scientific discoveries but then failed to follow up on them, so the torch of learning passed to the West. This is to read the empirical methodology and practical goals of modern science back into the intellectual methods and spiritual goals of the wisdom tradition. The goal was not to establish a fund of transmitted knowledge which other scientists could imitate and build upon and from which technologists could draw for practical ends. The goal was to discover the truth for oneself.

Rūmī sums up the difference between a *muḥaqqiq* and a *muqallid* – between someone who thinks for himself and someone who imitates others – in the following verses:

> A child on the path does not have the thought of men.
> His imagination cannot be compared with true *taḥqīq*.
> The thought of children is of nurses and milk,
> raisins and walnuts, crying and weeping.
> The *muqallid* is like a sick child,
> even if he offers subtle arguments and proofs.
> His profundity in proofs and objections
> drives him away from true insight.
> He takes the collyrium of his secret heart
> and uses it to offer rejoinders.[9]

Rūmī, then, speaks for the whole Islamic intellectual tradition when he says that no one can achieve true and real understanding until he stops imitating others and finds out for himself. The implication for the modern situation is clear: there can be no rehabilitation of Islamic thought unless Muslim thinkers put the *taḥqīq* of *tawḥīd* back at the center of their concerns.

4

Beyond Ideology

One of the many roles that a living intellectual tradition would play is to help people understand the nature of ideology, by which I mean any sort of sociopolitical program built on analyses of human nature that are deemed to be rational and scientific. Defined as such, ideology is rooted in the humanistic and secular theories that grew up in the Enlightenment. It does not include traditional religion, that is, premodern forms of religious thought, though it does include the various forms of politicized religion that are lumped together as "fundamentalism," given that they represent specific varieties of modern thought.[10]

Ideology provides the theoretical framework for practically all political and social thought in the modern world, so there is no escape from its influence. Nonetheless, the intellectual tradition may suggest some of the ways in which we as individuals can navigate past its shortcomings. Specifically, I have in mind three important goals of this tradition: breaking the shell of dogmatism, asserting the absoluteness of the Real, and resuscitating the mythic imagination.

THE OMNIPRESENCE OF TRANSMISSION

In any field of transmitted learning, experts have several important concerns. These include organizing and interpreting their knowledge and shoring up the reputation of those from whom knowledge is transmitted, that is, the "authorities." In the Islamic context, attempts to prove the reliability of transmitted knowledge are obvious in the activities of theologians and jurists, given that their whole enterprise builds on the transmission of the Qur'an and the Hadith. But the same need is present in all transmitted knowledge.

It should be obvious that the fundamental transmitted knowledge of any culture goes largely unquestioned. People receive it as part and parcel of their language, customs, techniques, artifacts, and everything they take as normal. Such knowledge is never simply religious. It may just as well be scientific or political or historical. If people are sure about something, this is because it goes unquestioned in their trusted circles. In their view, "Everybody knows that." We do not normally question the authority of those who establish the very structure of our categories of thought. Transmitted knowledge is woven into the fabric of our worldview, whatever that worldview may be.

Transmitted knowledge, then, is the type of knowledge that dominates over human culture, and modern culture is no exception. When we imagine that we know something, we have heard it from others. Nor can we claim that our own personal and experiential knowledge qualifies as intellectual, because we have received it from our sense organs, which are notoriously unreliable, and we have interpreted it in terms of the prevailing worldview.

In contexts where the authority of transmitted knowledge was sustained primarily by religious belief, there were few sources of authoritative transmitted knowledge, so there were relatively few categories of teachers. Nowadays, various systems of knowledge compete with each other with chains of transmission going back to the founding fathers. There is an enormous proliferation of privileged classes claiming to represent authoritative knowledge – scientists, engineers, doctors, psychiatrists, lawyers, physicists, neurosurgeons, Orientalists. No matter what we want to say about the reliability of such knowledge, for you and me it is transmitted. What gives us confidence in it – if we have any – is that we trust the authority of the source.

If transmitted knowledge is our ordinary, everyday sort of knowledge, intellectual knowledge is something quite different. Knowledge only qualifies as intellectual when knowers know it at the very root of their own intelligence and without any intermediary – not even imagination and cogitation. In the terminology of Islamic philosophy, this sort of knowledge was called "non-instrumental" (*ghayr ālī*). This is because it does not depend upon any of the "instruments" of the soul, the faculties and powers of the mind. It does not come from outside the self, nor does it derive from sense perception, imagination, cogitation, or intuition. It wells up from the deepest realm of intelligence, which is nothing but the divine spirit, the intellect at the root of the human *fiṭra*.

In short, the role of the intellectual tradition was to make first-hand knowledge available to those who wanted it. It was to show people the way to move beyond what they had been told. It was a path to discover the ultimate truths of the universe within the depths of one's own soul, the only place where truth can be found. This was the object of the quest. How many people reached the goal? Probably not very many. The point here is that the quest remained an ideal in Islamic society and that it kept aspiring philosophers and intellectuals focused on *tawḥīd* rather than *takthīr*.

BREAKING THE SHELL OF DOGMATISM

Let me now turn to the first of the three goals of the intellectual tradition mentioned earlier – overcoming dogmatism. By "dogmatism" I mean the claim put forth by teachers or thinkers or ideologues that everyone must adhere to a certain set of beliefs and practices as transmitted from their own trusted sources and interpreted by themselves. Dogmatism is no doubt a fact of life in all societies. In the Islamic context, the dogmatists were usually jurists and theologians, who claimed that all truth had been revealed in the Qur'an and that their own interpretation of that truth had to be accepted. In modern society, dogmatism is found among believers in every sort of god – religion, science, democracy, socialism, progress, freedom, development, and so on.

One of the results of the gradual weakening of the intellectual tradition over the course of Islamic history was the increasing tendency toward dogmatic closure, especially with the shaping of the juridical and theological schools. Nonetheless, we need to remember that the theologians and jurists, however narrow their perspective may have been, played the necessary role of preserving the transmitted knowledge upon which the religion depends. Moreover, when and if the theologian-jurists brought about dogmatic closure, they did so only in the sphere of transmitted knowledge, not in intellectual knowledge. Catechisms and polemics cannot hold people back from striving to achieve firsthand knowledge of God, the cosmos, and their own souls. The deep-rooted quest for wisdom that is innate to the human spirit cannot be blocked by rhetoric and threats. Certainly, it remained an open path in Islamic civilization. In the West, however, with the rise of science and secularism, the quest for wisdom was largely debunked, and our great and respected thinkers began talking about the death of God and the death of metaphysics that goes along with it. These notions have since become foundational in modern forms of transmitted knowledge.

Al-Ghazālī among others frequently attacks the dogmatic mentality. In doing so he explains that transmitted knowledge too often becomes a veil that prevents any attempt to achieve intellectual understanding. He writes, for example,

> The cause of the veil is that someone will learn the creed of the Sunnis and will learn the proofs for that as they are uttered in dialectics and debate. Then he will give his whole heart over to this and believe that there is no knowledge whatsoever beyond it. If something else enters his heart, he will say, "This disagrees with what I have heard, and whatever disagrees with it is false."
>
> It is impossible for someone like this ever to know the truth of affairs, for the belief learned by the common people is the mold of the truth, not the truth itself. Complete knowledge is for the realities to be unveiled from within the mold, like a kernel from the shell.[11]

The belief of the common people is precisely what they have received by way of transmitted knowledge. It is knowledge based on *taqlīd*, not *taḥqīq*. Only the latter gives access to *ḥaqq*, "the truth itself." This word, the root of the word *taḥqīq*, means not only truth, but also reality, rightness, appropriateness, worthiness, and duty. In Qur'anic usage, it sometimes carries a sense similar to our modern concept of "right." Nowadays its plural, *ḥuqūq*, is commonly used in talk of "human rights." Often forgotten, however, is that the Arabic word can just as well be translated as "responsibility." In the premodern discourse, rights and responsibilities were two sides of the same coin, both founded on the Absolute *ḥaqq* that is God.

When contemporary Muslim thinkers criticize *taqlīd*, the issue is always the interpretation of legal, social, and political teachings. To appreciate that it has nothing to do with *taḥqīq*, it is sufficient to note that they never attack *taqlīd* in all transmitted knowledge, only in the forms that they do not like. They themselves have taken what they know about Islamic history and society from others. Their criticism is addressed at the authority of those whose interpretation of Islamic law has come to be accepted. They are asking believers to stop imitating the old authorities and to start imitating the new authorities, who often seem to be themselves. They question the reliability of the transmitted knowledge that Muslims have been following for centuries. Most of them tell us that Islamic teachings have to be adapted to the times. Their basic argument, in other words, is that there are new forms of authoritative, transmitted knowledge that must now be imitated. This new transmitted knowledge has been established by contemporary theologians and jurists – now known as scientists, psychologists, biologists, sociologists, and critical theorists. The new authorities must be followed along with or instead of the old.

In the intellectual tradition, *taqlīd* was condemned in intellectual knowledge, not in transmitted knowledge or in the early stages of the quest for realization. In matters pertaining to social, legal, and other secondary affairs, *taqlīd* was considered appropriate, because transmission is precisely the source of such knowledge. We moderns have a rather different way of looking at things. We seem to think – or at least we act as if we think – that we should accept as given the popular consensus on the nature of the world, one that has been established by scientists, scholars, and the media. We ourselves, after all, lack the expertise. At the same time, we feel relatively free to be "creative" in our own thinking. We go about achieving creativity not by making contact with the transcendent source of creativity, which is the divine breath blown into the *fiṭra*, but by rebelling against the transmitted knowledge that forms the basis of law, religion, social order, and human relationships.

In short, *taḥqīq* demands not only knowing for oneself the First Truth and Absolute Reality, but also acting appropriately. The First *ḥaqq* delineates the *ḥuqūq* – human rights, duties, and responsibilities – by its very nature. Understanding these *ḥuqūq* requires conformity with them. *Taḥqīq* embraces both the cognitive act of knowing the *ḥuqūq* and the ethical responsibilities that follow upon the knowledge.

Ideology, in contrast, is built on the imitation of beliefs established by the fathers of modern thought, the prophets of modernity. These prophets in turn base their claims to authority on the scientific worldview established by the Enlightenment. From beginning to end, ideology demands belief in the authority of transmitted knowledge, not in truths that we have come to know for ourselves.

ASSERTING ABSOLUTENESS

If one goal of the intellectual tradition is to overcome dogmatic thinking by breaking the shell, finding the kernel, and knowing the True Reality for oneself, a second is to assert the absoluteness of the Real. This means to see all things in terms of their ultimate point of reference. The methodology of *taḥqīq* assumes that human intelligence is adequate to the Real and that the Real is one. The truth and reality of God and the universe – their *ḥaqq* – can be known; the rights of God, people, and other creatures – their *ḥuqūq* – can be discerned; and the appropriate and worthy response to truth and right can be put into practice.

By saying that "human intelligence is adequate to the Real," I do not mean to imply that the practitioners of *taḥqīq* ignored the insights provided by revelation in general and the Qur'an in particular. Certainly some theologians and jurists accused philosophers of denying God's messengers, or Sufis of considering themselves greater than the prophets. The basic reason for such criticism is obvious: the self-appointed defenders of the tradition tried to impose dogmatic closure on all believers, but the philosophers and Sufis wanted to know for themselves. They refused to rely on any knowledge that they had learned by way of hearsay, even if religious and social conventions maintained that the knowledge was true and reliable.

We must not forget that revelation addresses both intellectual and transmitted knowledge. The two domains are already highlighted in the two halves of the Shahadah. The first half addresses *tawḥīd*, the foundation of all intellectual knowledge, and the second half prophecy, the principle of transmitted, religious knowledge. The first half transcends history, because it simply asserts the nature of things. The second half – "Muhammad is God's messenger" – pertains to specific historical circumstances that can only be known by way of transmission.

Despite the dependence of the second half of the Shahadah on transmission, it raises questions about the nature of prophecy and revelation that are not contingent upon history and were considered accessible to intelligence without transmission. For example, what sort of human being is designated by the word "messenger"? Why should the authority of such a person be accepted? What is the difference between prophetic knowledge and merely human knowledge? What is the relationship between prophetic knowledge and ultimate human happiness?

The philosophers investigated these sorts of questions as intellectual rather than transmitted issues. They were not especially interested in the historical events surrounding Muhammad and other prophets, or in the details of the revealed scripture. Nor, in the early period, did they defend the graphic Qur'anic depictions of the afterlife as anything more than a rhetorical necessity. However, they were extremely interested in prophecy as the highest form of human perfection, and they were especially concerned with the immortality of the soul, which was to be achieved precisely through intellectual perfection.

For many of the theologians and jurists, the very act of asking questions about the second half of the Shahadah looked like unbelief. They wanted blind acceptance, without asking why. But the philosophers saw clearly that one cannot prove the authority of the Qur'an by calling on the Qur'an's authority. If we are talking about knowledge and not simply belief, then one must prove – without recourse to authority – that the Qur'an has authority. In order to do so, one must establish a necessary role for prophets in human history. If such a necessary role exists, it must pertain to human nature. It follows that the necessity of prophecy must be discoverable within human nature without transmission. If one does conclude that transmitted knowledge plays an important or necessary role, then one can take full and confident advantage of it.

Because the philosophers discussed the three principles of faith with little explicit reference to transmitted learning and much mention of Greek antecedents, some historians have found it easy to ignore the thoroughly Islamic character of their writings. Such historians have allied themselves with the Muslim critics who attacked the philosophers because their interpretations did not coincide with theological and dogmatic readings. Nonetheless, in a broad view, philosophy and theology were largely in agreement, especially if we compare their positions with the beliefs that inform most modern forms of scholarship, not to mention ideology.

My basic point here is that Muslim intellectuals saw themselves as investigating things in the context of the most fundamental insight of the Islamic tradition, and they did not see their efforts as opposed to the goals and purposes of the ulama. They accepted that the prophets came to remind people of *tawḥīd* and to teach them how to live in conformity with the One God. They also believed, however, that the vast majority of people had one path to follow, and that those drawn to intellectual pursuits had another.

From the standpoint of the intellectual tradition, there is no antagonism between intellectual and transmitted knowledge. One can perfectly well discover the truth of things for oneself and at the same time recognize the necessity of transmitted knowledge. The standpoint of transmitted knowledge, however, is quite different. If we reject the possibility of intellectual knowledge, we are forced to cling to the shell of knowledge, and the result will be dogmatic closure. Without understanding that the primary truths must be known for oneself and in oneself, we will choose to imitate others and accept hearsay as the basis for belief and action.

It should be obvious that in modern times, we live in a society that considers this sort of intellectual knowledge as an absurdity or an impossibility. As a result, there is always a feverish search for reliable transmitted knowledge, and this helps explain the mythic aura surrounding scientific discoveries. People believe that science alone is qualified to uncover the secrets of the universe, and not only that, they accept the discoveries as reliable truth, not realizing that they are asserting their belief in the authoritative knowledge of the priesthood of science. As for ideology, it always appeals to the gods Science and Reason as its justification, and it calls out to the human hunger for guidance and meaning, aiming to mobilize those who believe in scientific progress and utopia.

Among Muslims, the new transmitted knowledge of the Islamist movements rejects the transcendent, ahistorical hope in salvation of the premodern tradition and replaces it with impossible dreams of a perfect society. Muhammad Arkoun has been especially astute in explaining how ideology has become the theoretical foundation for all the political factions vying for power in Muslim countries. As he puts it, Islam has been turned into "an instrument of disguising behaviors, institutions, and cultural and scientific activities inspired by the very Western model that has been ideologically rejected."[12]

MYTHIC IMAGINATION

If two of the goals of the intellectual tradition are to overcome dogma and to assert the absoluteness of the Real, a third is to recognize the proper role of myth in human understanding and, if necessary, to revitalize mythic discourse. The Enlightenment succeeded in establishing the supremacy of instrumental rationality by rejecting the cognitive significance of myth and symbol, which are characteristic of scripture and much of religious discourse. The invisible realms to which the traditional language referred – God, the angels, life after death, human perfection – were seen as unintelligible and meaningless, because they could not be addressed by the empirical methodologies of instrumental reason.

On the Islamic side, the tendency of both theology and jurisprudence was to devalue the symbolic content of the religious teachings. Jurisprudence was interested in providing concrete guidelines for human behavior, and theology wanted to defend rationalistic dogmas abstracted from the symbolic language of the Qur'an. But these approaches were by no means adopted by the intellectual tradition. Sufis, and to a lesser degree philosophers, looked upon the signs and symbols of the Qur'an as a means to open up the soul to the presence of the Real in all things.

Modern scholarship has gone a long way toward rediscovering the role of myth and symbol in premodern civilizations and cultures. But modernity in general lacks the resources for understanding the real significance of what was going on. The reason for this is simply that it has failed to come up with a proper metaphysics, cosmology, and spiritual anthropology. By "proper" I mean "dealing with the *ḥaqq* of things," not simply with things as they are described in the transmitted learning of an ideological and scientistic age. Contemporary academic sciences have in fact been constrained by the dogmatism of transmitted sciences such as physics, biology, psychology, and sociology. As a result, theorists have placed arbitrary limits on human possibility.

The real danger of instrumental rationality lies in the dogmatic and absolutizing claims made by its supporters. Instrumental rationality must play a certain role in any society, to be sure, but when it plays the dominant role, the traditional teachings about human nature are necessarily obscured. In the extreme case of the modern West, scientific knowledge itself has usurped the role of myth and symbol. This helps explain why scientism pervades the modern imagination, so much so that most people – religious people included – simply take its assumptions for granted. Scientism is a rationalizing ideology that has all the persuasive powers of technology, education, and the media to back it up. It provides the de facto theology for the civil religion of modernity. The many contemporary thinkers who criticize it have no effect on the thinking and preaching of our own home-grown theologians and jurists – the scientists, technocrats, and journalists who have long since established a new set of myths and symbols to drive the modern world.

Because of the omnipresence of scientism, few people have any sense of the full-bodied truth and total coherence of premodern worldviews, which established delicate balances between mythic imagination and rational inquiry. In the Islamic context, no one has analyzed this balance with more subtlety than the enormously influential thirteenth-century jurist, theologian, philosopher, and Sufi Ibn 'Arabī. Let me summarize what he has to say on this vital issue.

Ibn 'Arabī maintains that we must see myth and reason as coexisting in harmony. The Real necessarily appears dichotomously to contingent beings. God is both creator and destroyer, both merciful and wrathful. Any analysis of the divine attributes shows that they must be understood both positively and negatively, both in terms of transcendence and in terms of immanence. The reason for this is simply that in itself, the Real is both absent from and present with everything in the universe.

Human beings, made in God's image, have a unique relationship with both God and the cosmos. This gives them the ability to grasp, understand, and realize God in both his distance and his nearness. Ibn 'Arabī calls the faculty of understanding God as distant "reason" (*'aql*) and the faculty of seeing God as near "imagination" (*khayāl*). What I have been calling "intelligence" or "intellect," he calls "the heart" (*qalb*), an important Qur'anic term that designates the synthetic, spiritual nature of human awareness.

If the heart is to perceive the Word of God resounding in itself, and if it is to intensify its own spiritual instinct, it must open what Ibn 'Arabī calls its "two eyes" – the eye of reason and the eye of imagination, or discursive thought and mythic vision. Only the fully realized heart can grasp the symbolic significance of revelation, because neither reason nor imagination on its own can see the fullness of the *ḥuqūq* – the truths, realities, rights, and responsibilities – established by the Absolute *ḥaqq*.

In Ibn 'Arabī's reading of the Islamic tradition, the eye of reason is the characteristic tool of the theologians and jurists. It is inadequate because it can only see God as transcendent. It recognizes that God cannot be known in himself, so it describes him as totally apart from every created thing and every quality. Left to its own devices, discursive reason will eventually reject the messages of the prophets – which are primarily anthropomorphic and mythic – and refuse to acknowledge that anything positive can be said about God.

In other words, excessive stress on rational thought pushes the divine into total transcendence. When this process is not kept in balance with the eye of myth and imagination, rational analysis eventually makes "the hypothesis of God" extraneous to rigorous, critical thinking. We see this process taking place in the mainstream development of Western thought. The end result is a scientific rationality completely oblivious to the *ḥuqūq* of God, the world, and the human soul. Excessive dependence on reason leads to agnosticism and atheism.

For its part, the eye of imagination sees God as immanent. It recognizes God's signs and marks in all things. It perceives the universe as the theatre of divine significance, infused with intelligent and intelligible light. It finds God's names and attributes manifest everywhere in the world and the soul, and it describes God in the positive terms supplied by revelation and the natural realm. This is to say that the eye of imagination feeds on myth and symbol, and it sees things not simply as signs and pointers to God, but as the actual presence of the Real. Left to its own devices, however, it will divinize the world and its productions and fall into *takthīr*, the assertion of many gods.

In Ibn 'Arabī's view, the heart is the unitary awareness at the root of the human selfhood. It is identical with the divine spirit that God blew into the clay of Adam, but it needs to be recovered, cultivated, and actualized. The goal of realization is to find the *ḥaqq* of the heart, the *ḥaqq* of God, and the *ḥaqq* of all creatures, and then to act according to all these *ḥaqq* s. No *taḥqīq* is possible unless one sees with both eyes, recognizing God in both his transcendence and his immanence, both his absoluteness and his infinity.

The heart, which is no different from realized intelligence, must employ the critical powers of reason to prevent associating other gods with God, or to avoid turning relative things into absolutes. But, if intelligence needs to employ reason correctly, it also needs to make proper use of imagination. It must undertake the mythic task of seeing everything as a sign and symbol of the divine. It must behold every creature as a "face" (*wajh*) of God and recognize that everything in the universe has a *ḥaqq* bestowed upon it by its Creator. It must keep the symbolic significance of things alive and respond properly to the living presence of God in the world. Only this attitude can allow people to respect the rights not only of God and other human beings, but also of the natural realm. When people fail to see the divine face wherever they look, they fall either into the one-sided transcendentalism that is characteristic of religious fundamentalism or the atheism and agnosticism that are characteristic of secular and scientific fundamentalism.

SELF-UNDERSTANDING

If the Islamic intellectual tradition has any help to offer to the modern predicament, it seems to me that it lies in the call to recover *for ourselves* – each of us individually – a proper understanding of our own nature. Otherwise, dogmatism and ideology cannot be avoided. The fundamental insight of the tradition is that in order to know the proper way of acting in the world and living out our human embodiment, we must know what the world signifies to us. In order to know the significance of things, we must know our own nature and our own proper destiny. In order to know our own nature, we must know the self that knows.

The point that is typically forgotten in discussions of who we are is that we cannot know the knowing self as object, only as subject. We cannot truly know ourselves except when object and subject are indistinguishable. The unity of knower and known, of self and world, of man and God, is the ultimate insight of *tawḥīd*. It is this alone that gives human beings the ability to see things as they truly are, to recognize the *ḥuqūq* of God, people, and things, and to act properly in response to the rights of God and the rights of man.

Offering a critique of dogmatism and ideology is a necessary first step if we are to recover a proper understanding of human nature. But proper understanding demands recognizing that the human self is grounded in a trans-historical intelligence and ultimately in Absolute Reality. As long as scientists and scholars persist in ignoring the fact that the soul cannot know the truth of things by standing on someone else's shoulders, there will be no escape from dogmatism, which is grounded in imitation and turns transmitted information into absolutes. Until it is recognized that the only dependable and real knowledge is awareness of the First Real, there will be no escape from an ever more polarized world of ideological conflict.

5

The Unseen Men

Thirty-some years ago someone told me about a lecture that had recently been given by Seyyed Hossein Nasr. During the question-and-answer period, the great Orientalist Gustave von Grunebaum remarked that Nasr's talk presupposed a power structure. What was it? Nasr replied with a sparkle in his eyes, "The *rijāl al-ghayb*," and von Grunebaum along with those who caught the reference laughed. Like all good jokes, this one has an element of truth in it – mythic truth, no doubt – but it certainly helps explain the voice of authority that often surfaces in Nasr's writings.

The term *rijāl al-ghayb* means literally "the men of the Unseen." In Sufi lore it refers to those human beings who live consciously in the spiritual world while governing the visible world as God's representatives. Although seldom recognized by others, they alone fulfill the cosmic function of human beings.

God created the universe, as the hadith puts it, "in order to be known." Among all creatures, only human beings have the capacity to know God in his full amplitude and grandeur. In their historical actuality, human beings are indefinitely diverse, and their diversity pertains to every modality of being and knowledge. It follows that some people are better at knowing God than others, just as some people are better at football than others. From the Sufi perspective, knowing God has relatively little to do with rational acumen, and much to do with God's gifts to those whom he chooses as his friends (*walī* or "friend" being the term that is commonly translated into English as "saint"). The Prophet reported that God says, "My friends are under My cloak – no one knows them but I." These unknown friends are precisely the Men of the Unseen, whether they be male or female (the Arabic word for "man" here has connotations not unlike those that allowed Latin *vir* or "man" to give rise to the word *virtue*).

According to some accounts, the Unseen Men can be divided into two sorts. One sort, known as the Men of Number (*rijāl al-ʿadad*), fill a static, ever-present hierarchy, their number never changing (some say it is 124,000, like the prophets from Adam down to Muhammad). Their chief is the Pole, who is the axis around whom the world revolves and the most perfect human being of the era. Outwardly, the Pole may be an ordinary and unremarkable person, but inwardly, as the texts put it, "He holds the reins of affairs in his hands." When the Pole dies, God replaces him with one of the two Imams, who had been the Pole's viziers, and he replaces the missing Imam with one of the four Pegs. Below the Pegs stand the seven Substitutes, and below them the twelve Principals. Among the Men of Number, one manifests the perfections of the angel Seraphiel, three the perfections of Michael, five those of Gabriel, seven those of the prophet Abraham, forty those of Noah, and three hundred those of Adam. The ranks of the Men of Number are constantly replenished as people pass on to the next world.

As for the second sort of Unseen Men, their number is not fixed, and they play a variety of roles according to circumstances. Most of them fall under the ruling authority of the Pole, but one group, known as "the Solitaries," stand outside his realm.

It is not clear how literally these reports are meant to be taken. No matter how we understand them, however, they speak eloquently of the intimate relationship that the intellectual tradition saw between cosmos and soul. This understanding of human nature underlies Seyyed Hossein Nasr's writings and helps differentiate his perspective from the typical historical or Orientalist approaches, which tend to provide brief and superficial overviews of Islamic theology and brief descriptions of the duties and obligations imposed on believers by the Shariah; then they quickly get down to the "real" business of describing the historical vicissitudes of the Muslim community.

Nasr often speaks of the loss of the traditional Islamic worldview and the havoc wreaked on the Muslim mind by scientific theories about the universe. As he points out, and as is obvious to those familiar with the contemporary situation, we are dealing with two diametrically opposed ways of looking at reality, even if many contemporary Muslims see no contradiction between belief in the Islamic God and belief in the objective status of scientific facts. Throughout the Muslim community, two basic groups of thinkers are found. One group, constantly becoming smaller, lives more or less in the traditional worldview. The other, ever on the increase, is led by engineers, doctors, and other professionals trained on the Western model.

These two groups do not speak the same language, and neither has any real idea of what the other is talking about. So utterly self-evident is the nature of the world to each group that they cannot imagine any other way of seeing it. The fact that they do not understand each other helps explain why contemporary Muslim preachers can exhort the young to study science and engineering, relying on prophetic sayings such as "Seek knowledge even unto China."

They speak of science here by using the term *ʿilm* or "knowledge," always recognized as the backbone of Islam, and they have no idea that science is driven by the worldview of *takthīr* or that it has starkly different goals and implications from Islamic knowledge, whether of the transmitted or the intellectual sort.

In several of his works, Nasr has explained the main principles of the traditional Islamic worldview. Here I will try to reformulate certain aspects of this worldview in a language as unencumbered by technical Islamic terminology as I can manage. My aim is to bring out the basic ideas on human nature underlying Nasr's writings, and especially his evaluation of modern thought. I offer one person's opinion that his interpretation of the contemporary implications of Islamic thought are firmly grounded in the tradition, much more so than many of his critics would like to acknowledge. The fact that he does not always cite Muslim authorities, but instead is likely to refer to Frithjof Schuon or Ananda Coomaraswamy, cannot be taken as evidence that his views do not have the Islamic support that he claims. He is not speaking as a preacher interested in bolstering his arguments by quoting the revered names, but rather as a philosopher who has found some of the clearest expositions of his own intellectual vision in contemporary authors.

Nasr, of course, does not write only about Islam, but also about other religions as well. Like Schuon and Coomaraswamy, he claims universal validity for a point of view that he and they usually call "traditional" and that observers have often called "traditionalist" or "perennialist." This perspective asserts that human beings at all times and in all places have recognized the reality of one unique principle and received guidance from it on various levels. What makes them human is not the peculiar biological, social, and historical constraints placed on the species, but the fact that they have been given access to the Infinite, the Absolute. This access is *given* to people, which is to say that it comes from the other side and cannot be reached by self-motivated efforts. This explains the necessity of prophets, avataras, buddhas, sages, shamans, and so on, and why the guidance must be transmitted from generation to generation.

Nasr and the traditionalists never take the patently absurd position that all claims to suprahuman guidance are true, nor do they say that all forms of revealed guidance will lead to the same "place." Evil and misguidance play important roles in the human situation. Nasr does offer some general principles as to how truth is to be discerned from falsehood and right from wrong, but by and large he leaves the assessment of specific teachings to the traditions within which they are offered. What is important for him is the principle of the universality of the guidance that comes from the Absolute and the fact that it is always available.

SUFISM

One of Nasr's subtexts is the relevance of Sufism to the contemporary situation and the catastrophic results that modern-day Muslims suffer by ignoring or rejecting it. For a great variety of reasons, people become suspicious at the mention of Sufism. In contemporary America, it is often associated with gullibility, sentimentality, and New Ageism. In the Islamic world over the past century, many Muslims have taken Sufism as a demonic presence that must be driven out if Islam is to enter the modern world, and today it is anathema to fundamentalists.

The fact is that relatively few modern-day Muslims have any idea of the historical role that Sufism has played, even though they are likely to have strong opinions on the topic. A colleague who teaches at Harvard recounts with amusement that a young Egyptian studying at MIT took a course with him on al-Ghazālī, who has universally been recognized as one of the greatest masters of the Islamic sciences and who is credited with authoritatively establishing the central role of Sufism in Islam. At the end of the semester, the student submitted a paper beginning with the sentence, "Islamic *taṣawwuf* does not exist" (*taṣawwuf* being the Arabic term for "Sufism"). This opinion, despite its incoherence, is widely held among Muslims, and the historical record is considered of no account.

Those Muslims who consider Sufism alien to Islam often draw support from the works of the early Orientalists, who saw it as a clear example of borrowing from other religions (after all, they imply, the Sufis were loving, open-minded, and well-intentioned people, so they could hardly have been real Muslims). Despite the fact that fundamentalists attack Western studies of Islam generally and Orientalism in particular, they are happy to accept this untenable theory of Sufism's origins.

Even specialists in fields like Religious Studies or Islamic Studies will sometimes remark, "Oh, but he's a Sufi," meaning, "You know, you do not have to take him seriously, because he's a mystic," or, "Sufism really has nothing to do with Islam, so don't pay attention to him." Yet for Nasr, and for the grand authorities like al-Ghazālī, the diverse beliefs, practices, and institutions of Islam that are apparent to outside observers make up Islam's body, and Sufism provides its life-giving spirit. From this standpoint, Muslim modernists and fundamentalists, who violently reject the Sufi tradition, are trying to breathe new life into Islam's body, and this life can only be drawn from alien sources. The discussion here, of course, is not about the history of the word *ṣūfī* (and its derivatives), since the term came into regular use only in the third/ninth century, but about what Nasr and many of the great authorities of the past have understood by the term when they employ it.

Although Nasr has written eloquently and persuasively about Sufism's centrality to the Islamic tradition, he cannot repeat these remarks in everything he writes, and even if he could, many observers reject this understanding of Sufism's role in Islam, so they feel no need to consider his position. Nasr has not necessarily helped his case by describing Sufism as "Islamic esoterism." In this he is presumably following Schuon (and to a lesser degree, Henry Corbin). Schuon has written voluminously, employing the esoteric/exoteric dichotomy as a key conceptual tool for understanding religion. However, not many English-speaking scholars have followed this practice, partly because few specialists have found it helpful in dealing with the actual texts.

One of the problems with the word *esoteric* is that, no matter how carefully terms may be defined, negative connotations cannot be avoided. The word is suspect by its very aura, and little can be done about it. From a linguistic point of view, one of its disadvantages is its high degree of abstraction, which results in a constricted semantic field that does not allow it to embrace the vast diversity of phenomena that have always been associated with Sufism. The restricted field becomes obvious if we compare the English word *esoteric* with the Arabic word *bāṭin* (or *bāṭinī*) of which it is sometimes said to be the translation. The two terms may indeed be employed in parallel ways on occasion, but *bāṭin* (which derives from the term *baṭn*, meaning "innards") has a concrete meaning and vast possibilities for metaphorical use. In other words, the basic meaning of *bāṭin* is "inner" or "inward," not "esoteric."

If it is said that Sufism emphasizes the more "inward" teachings of Islam, few scholars would object. The point is simply that Sufism's perspective contrasts with that of disciplines like jurisprudence and Kalam, which emphasize the more literal and socially oriented teachings. The terms *inward* and *outward* are broad and inclusive enough so that everyone will understand an appropriate meaning without being drawn into irrelevant questions, such as the elitism and occultism that are typically associated with esoterism.

Both *esoterism* and *exoterism* introduce nuances and connotations that are not present in the Arabic terminology. Once Nasr and others use the words, it makes sense to criticize them for being esoterists or for supporting the views of contemporary occultists – and people quite sympathetic to Sufism have done so.

COSMOS AND SOUL

The cosmic role of human beings lies in the background of many of the criticisms that Nasr levels at the scientific worldview. The notion of the "Men of the Unseen" is one way of expressing some of the tradition's fundamental insights, and the ideas lying behind it can help us understand why Nasr stands where he stands.

In the traditional, broad-based Islamic view of things, one cannot disengage the study of the soul from cosmology. Of course, everyone recognizes that premodern Islamic psychology has much to do with theology, since Islam agrees with the Judeo-Christian tradition in holding that man was created in the divine image. But nowadays, the cosmic dimension of Islamic psychology is difficult to understand and easy to ignore, not least because cosmology in the West has long since been delivered over to natural science.

Most contemporary Muslim thinkers, in their eagerness to prove Islam's respectability in modern terms, have ignored or attacked those Islamic teachings on the cosmos and human beings that are difficult to reconcile with the contemporary worldview. They do so by ignoring the commentarial tradition and interpreting the Qur'an in terms of their own immersion in ideology and scientism. Others have appealed to Kalam, the most rationalistic form of Islamic theology and the least concerned with the nature of God's ontological relationship with the universe. Kalam is polemical and voluntarist, devoted to nit-picking attacks on any form of thought that is deemed to threaten God's absolute legal authority. It asserts

God's radical transcendence and argues vehemently for human responsibility before the revealed law.

Educated Muslims generally see things in terms of the worldview that has informed the Western tradition since the beginning of the modern period. This worldview is grounded in what Nasr calls a "sensualist and empirical epistemology," and its net result has been the reification and objectification of the cosmos. The world and all its contents, including human beings in most of their roles, have been turned into isolated objects standing in ontological, spiritual, and moral vacuums.

In the West, ecologists of various stripes have attempted to show the short-sightedness of current conceptions of the world, usually in terms of an enlightened self-interest. Some have gone so far as to propose alternative cosmologies, but these are almost always "scientific" in that they take for granted the necessity for empirical verification and the nonexistence of any truly transcendent dimension to reality. Or, they simply accept the theories of modern science at face value, and try to construct a new, scientific mythology, which will somehow restore wonder and respect to human observers of the world. Still others have recognized the need to recapture transcendence and, in trying to do so, have cobbled together diverse notions from science and various traditional worldviews with the hope that they can come up with a softened and sensitized scientific mind-set.

Nasr's critique of scientism and technology is rooted in the understanding that science, standing on its own, cannot conceive of what it means to be human. Many serious scientists, at least, are aware of these limitations, but not the scientific popularizers, who have the most effect on how people perceive the world. As long as the truncated worldview of scientism remains the arbiter, no opening to the Infinite is possible. At best, people will devise an ersatz cosmology that hardly lets them see beyond the horizons of popular culture.

NAMING REALITY

There are many versions of Islamic cosmology, few of which have been studied in modern times. Common to all of them is *tawḥīd*, the axiom that there is one supreme principle, an ultimately unnamable and unknowable principle, and that everything in existence appears from it and returns to it. Different schools of thought discuss the modality of appearance employing a variety of terminology, such as divine fiat, creation, and emanation. Once we recognize that the ultimate principle is there, it can be given various names, with the reservation that the names do not really help us to understand the named in itself, in its very essence (*kunh dhātihi*). Nonetheless, naming the principle is a necessary stage in coming to understand its implications for the human situation. And truly efficacious naming, that is, efficacious in terms of the full reality of what it means to be human, comes from the principle itself.

Naming has repercussions by nature. When we name something, we situate it in a pre-existent view of reality that allows the name to have meaning. We deal with things in terms of the names that we give to them. If we name something a "chair," we sit on it, and if we name it "firewood," we burn it. The Islamic tradition – like other traditions – names the world and its diverse contents in ways that let people see the function and role of human beings. This is conceived of in terms of the divine compassion that has brought the universe into being in the first place.

The Qur'an tells us that God taught Adam all the names, a verse that epitomizes Islam's theology, cosmology, and spiritual psychology. It alerts us to the three basic realities that must be taken into account if we are to understand the nature of things – God, cosmos, and soul. God taught the names of all of these to human beings at their origin. The names were in no way divorced from their meanings, that is, the realities named by the names. Rather, the names were the perfect expressions of Adam's realization of all knowledge in the depth of his soul. This is Rūmī's point in these verses:

> The father of mankind, the lord of "He taught the names,"
> had a hundred thousand sciences in every vein.
> The name of everything as it is until its end
> was given to his soul.
> Whatever title God gave never changed.
> The one He called "quick" did not become slow. ...
> For us, the name of each thing is its appearance;
> for the Creator, the name of each is its inner mystery. ...
> Adam's eye saw with the Pure Light,
> so the spirit and mystery of the names became plain to him.[13]

As the first prophet, Adam is the primordial recipient of divine guidance and the leader of all his children on the road to salvation and realization. If his children are to deal with the cosmos properly and appropriately – according to the *ḥaqq* of things – and if they are to actualize the fullness of their own nature (*taḥqīq*), they need to understand the names revealed to their father and act accordingly.

Human beings will always name things, because they are by definition "talking animals" (*ḥayawān nāṭiq*). This expression is usually translated as "rational animals," in keeping with the way the ancient Greek expression entered English, but the Arabic *nāṭiq* or "talking" highlights an important nuance of the Greek. Human rationality is articulate, uttered, spoken; and proper human speech is intelligent and rational. In the Islamic worldview, the full actualization of this spoken, articulate rationality presupposes knowledge of the real names of things. Knowing the real names means knowing things in the context of God's knowledge of them, which only comes to us when he himself names them for us.

If people fail to name things under the wing of divine guidance, they will name them as they see fit. There is no possible way, however, for them to know the real names of things without assistance from the divine Namer, because the real names are the realities of things in the divine mind. God gives existence to the things according to their names, and understanding their real names is the key to understanding cosmos and soul. A worldview that leaves out the divine dimension will necessarily deal with inadequate names, if not misnomers. The net result of misguided naming will be disaster for those who employ the names, if not for humanity as a whole – a "disaster" that is understood in terms of the full extension of the human realm, not just the world this side of death.

To understand why modern, scientific cosmology appears to the intellectual tradition as enormously truncated, it is sufficient to meditate on the names that science gives to the really significant things, the mysterious principles or realities that determine the configuration of the real world. What happens when the important names are quasars, quarks, muons, black holes, and big bangs? What is the psychological and spiritual fruit of naming ultimate things with mathematical formulae?

The basic characteristic of the mathematics that is nowadays deemed capable of expressing the nature of things with authority is its abstraction, its abstruseness, its reconditeness – the fact that only a tiny elite are able to grasp its significance and explain it to the commoners. The more the experts learn of the ultimate mysteries of the scientific universe and reduce it to mathematical formulae, the more they find that it is impersonal, unintelligible (to the commoners), and arbitrary. The cosmos, the hard-nosed scientists tell us, is inhuman, and human beings are an oddity, a cosmic accident. The trickle-down effect of this worldview is palpable in modern culture. Appleyard sums it up nicely in his analysis of "liberal man," the enlightened individual whom our most progressive thinkers hold up as the ideal:

> Unable to create a solidity for himself, liberal man lapses into a form of spiritual fatigue, a state of apathy in which he decides such wider, grander questions are hardly worth addressing. The symptoms of this lethargy are all about us. The pessimism, anguish, skepticism and despair of so much of twentieth-century art and literature are expressions of the fact that there is nothing "big" worth talking about anymore, there is no meaning to be elucidated.[14]

Islamic cosmology begins with the knowledge that the universe holds the keys to the immortality of our souls. It views the cosmos as instilled with meaning and purpose. It names the One Origin of the cosmos with a variety of names derived from the divine self-naming. None of these names is abstract or inhuman. The Islamic God is anthropomorphic, because the Islamic human is theomorphic. If God is understood in man's image, it is because man was created in God's image. Unless God is understood in human terms, a yawning gap will remain between the ultimate and the here and now. *Re-ligio* or "tying back" to God is impossible without images of God and imagining God.

People need to take an active role in tying themselves back to God, and they can only do so in terms of themselves and their own understanding. They can understand only what they are. If they do not display the traces of the divine in some way, they cannot tie themselves back to the divinity. People who live in a traditional, anthropomorphic universe will necessarily deal with it in human terms. Those who live in an abstract universe will deal with things and others as abstractions. Those who live in a mechanistic universe will treat everything as a machine. Those who find the universe cold and uncaring will reciprocate.

It is true that Kalam and some forms of Islamic philosophy assert God's absolute transcendence and claim that the names of God should not be understood in human terms. This perspective is necessary, because it helps preserve the understanding that things begin with God, not with us. As Ibn 'Arabī frequently tells us, the proper role of rational thought is precisely to assert and maintain the transcendence of the One. But the mythic imagination also has its rights, for the Real is in fact "with you wherever you are," as the Qur'an puts it (57:4). The anthropomorphism of the intellectual tradition results from seeing with "both eyes," the eye of reason and the eye of imagination. It is not the crude sort that we hear about in unsympathetic accounts of polytheistic worldviews, but rather the recognition of the mercy, goodness, and wisdom that pervade reality, whether or not we grasp how these qualities are present in any given circumstance.

Although the Qur'an's depiction of God is far from that of polytheistic myth (in the Hindu or Greek sense), it is certainly polynomial. The Qur'anic names of God, enacted and performed in the diverse modalities of ritual and praxis, determine the traditional Muslim mind-set far more than the abstractions of the Kalam experts or the rules and regulations of the legal scholars. To the extent that Muslims put their religion into practice and assimilate the Qur'an's teachings, they cannot fail to see God's wisdom in the signs and phenomena of the universe and the self, just as they see it in the signs and phenomena that are the verses of the Qur'an.

Muslim praxis is studded with the divine names. Every significant act begins with a formula that epitomizes more than any other the Muslim understanding of God and his relationship with his creation: "In the name of God, the All-merciful, the Ever-merciful." God deals with the universe in terms of his own names, and his primary names assert his universal mercy and compassion. Every prayer, every supplication, every act of remembrance (*dhikr*), is highlighted by divine names. And every rational attempt to understand these names is propelled by the intuition that God lies infinitely beyond human conceptualization.

God gives, and he takes away. He gives the names through his revelations, and he takes away our understanding of them through our attempts to understand them. The more we try to grasp their significance, the more they turn us back to the unknowability of God in himself. These are the two movements of the divine and the human – descent and ascent, origin and return, revelation and concealment, disclosure and curtaining. They mark a creative dynamic in Islamic culture that has totally disappeared in the monolithic thinking of Muslim modernists and ideologues.

Muslims who practice the Prophet's Sunnah and live in the Qur'anic universe cannot help but think of cosmos and soul in terms of the revealed divine names. These are not strictly personal names, nor are they impersonal. God is alive, knowing, desiring, powerful, speaking, hearing, seeing, creator, life-giver, death-giver, forgiving, pardoning, avenger, bestower, withholder, and so on. The names of the ultimate reality establish the meaning and significance of what people encounter in the signs.

The universe is imbued with purpose, and the individual instances of its purpose become clear when situations are understood in terms of the divine attributes that become manifest through them. Not that this is easy – how can we be sure if an instance of our happiness displays God's mercy or his wrath, his compassion or his vengeance? We have no way of knowing the final outcome of affairs.

Traditional Muslims are confident, however, that things will work out for the best, no matter how badly they may go in any given situation. "In the name of God, the All-merciful, the Ever-merciful" announces all phenomena of the universe. The Qur'an says that God's mercy "embraces all things" (7:156), and the Prophet added a subtlety to the point with his famous saying, "God's mercy takes precedence over His wrath." This is an ontological and cosmic precedence, and it means that all is well in the divine scheme of things. It follows that, as the Prophet put it, "The believer is fine in every situation."

The Qur'an repeatedly commands the believers to have trust in God, and the attitude of trust in God's mercy infuses the traditional worldview. Ideologues and fundamentalists ask Muslims to trust instead in utopian dreaming, military technology, and centrally planned, coercive applications of the Shariah. Then alone, they tell us, will Islam be put back in the driver's seat of history where it belongs. They never question the legitimacy of the impersonal view of reality that has allowed science and engineering to dominate people's understanding of the world in the first place.

THE ONE AND THE MANY

Although the One God in himself cannot be known, his manifestations cannot be avoided, so much so that it can be said that from a certain point of view, nothing can be known but the One. However, knowledge of the One's infinitely diverse manifestations is infinitely diverse, which is to say that God is known through *takthīr* as well as *tawḥīd*. Knowledge that clings to the data of sense perception (whether or not this is mediated through instruments) is limited to the surface, the outward, the superficial, the skin – all these terms understood as metaphors, not as literal, scientific designations. The One can only be truly known inasmuch as it names itself, and these divinely taught names have everything to do with our understanding of how the universe comes into existence.

A typical listing of the divine names that generate the cosmos begins with alive, knowing, desiring, and powerful. Among these, alive is especially interesting. When Sufi theoreticians like Ibn 'Arabī explain the nature of the divine life, they are likely to employ the term *wujūd*, which is typically translated as "existence" or "being." The Arabic word, however, also means finding, awareness, consciousness, and joy. There can be no such thing as an inanimate and unaware *wujūd*. It makes no sense to think of *wujūd* simply as "existence," the fact of being there, some sort of cold inanimateness within which life and joy and love are cosmic accidents. An implicit if not explicit side to the use of this term is that God's own life, awareness, and consciousness course through everything that exists, though his attributes display themselves most clearly in what we call "living things" – plants, animals, and human beings.

The single, supreme Principle manifests itself through multiplicity, but this is an ordered and hierarchical multiplicity, one that begins with twoness and gradually differentiates itself into various cosmic levels. Twoness is an especially important notion in cosmological thinking, because it allows us to conceive of a world along with the supreme One. The duality that appears when we conceptualize the world next to God colors all the relationships between the One and the many and has repercussions throughout the cosmos.

For many cosmologists, the basic duality of God and the world gives rise to two complementary points of view. From one standpoint, God is utterly real and the world utterly unreal; from another standpoint, the world has a relative reality (when compared to pure nonexistence), and this reality can only derive from God. Inasmuch as we emphasize God's reality and the world's unreality, we conceive of God and the world in terms of insuperable otherness. Inasmuch as we conceive of God as giving rise to the world through his activity and attributes, we conceive of God and the world in terms of unfathomable sameness. In other words, God is both transcendent and immanent (or, as I prefer to translate the Arabic terms, both "incomparable" with all things and "similar" to them).

In terms of God's transcendence, the world is nothing. In terms of his immanence, it is something, because it displays the attributes and qualities that he bestows upon it. True life and consciousness belong to God alone, and everything else is strictly dead. But once we note the divine life in the cosmic signs, we see that everything is alive and aware to some degree.

The vertical duality that differentiates God from the world gives rise to the understanding of a horizontal duality *in divinis* – a duality sometimes referred to in Qur'anic terms as God's "two hands." Inasmuch as God is distant, transcendent, and incomparable, he is conceived of in the guise of the names of majesty; inasmuch as he is distant, immanent, and similar, he is conceived of through the names of beauty. Ultimately, "God's mercy takes precedence over His wrath," because beauty and gentleness pertain to God's fundamental reality, but majesty and severity pertain to him only when he is understood as distant from his creatures. Creatures, however, have no reality of their own through which to remain distant from God, so they can only stay in nearness and sameness, despite the vagaries of time and the unfolding of the diverse possibilities of otherness.

THE LIVING UNIVERSE

From the point of view of Islamic cosmology, what we call "science" is a reading of the universe that ignores all but the most insignificant meanings that the cosmos has to offer. When the universe is named by names that apply primarily to dead things or to machines or to impersonal processes, we will understand it in terms of death and mechanism and impersonal process. We will necessarily miss the significance of the life, mercy, and awareness that suffuse its every atom.

A Sufi axiom holds that "*Wujūd* descends with its soldiers." *Wujūd* here designates not only the Being of God, but also his finding, consciousness, awareness, and joy. It is God's life in himself, which is then reflected in diverse degrees in all things in the universe. It leaves its traces in the cosmos when it "descends," that is, when God creates the universe, thereby bestowing reality upon it. In God, it is pure, which is to say that God is simply *wujūd*, nothing else – pure being, sheer finding, undiluted consciousness, utter bliss, infinitely effulgent light. When God creates the universe, he does so by dimming the light in keeping with his infinite wisdom. Wherever anything finds and is found, this is nothing but the refracted light of *wujūd*.

Wujūd's "soldiers" are its attributes, the qualities by which it is named in its manifestation. We come to know them when God names himself by them, and he does so in scripture, in the cosmos, and in our own souls. Through studying any of these, we come to understand that he is alive, knowing, powerful, merciful, wise, and all the rest. Every name leaves its traces in everything in the universe, even if we fail to perceive them. The names are omnipresent, because *wujūd* is omnipresent, failing which the things would not be found.

Just as God is absent from all things because of his transcendence, absoluteness, and incomparability, so also he is present in all things because of his immanence, infinity, and similarity. Because of his reality (*ḥaqq*) in face of our unreality, it is he alone who establishes our reality and the realities and rights (*ḥuqūq*) of all things. Because of the relative reality that we gain, we have the responsibility (*ḥaqq*) to respond to the rights and realities that we face. From one point of view, the realities that we face are ontological and cannot be avoided, for we are God's productions, totally passive in the hands of his creativity. But the relative fullness of God's presence in his human image bestows upon people a certain freedom, and it is this that results in the rights of the soul and the rights of the other, the moral and spiritual responsibilities that give meaning and direction to our world. We have no choice but to try to live up to the divine attributes found in ourselves and the cosmos. The King and his soldiers are present in all things, in all "objects."

There can be no moral vacuums, no hideouts for "pure objectivity" and "scientific disinterest," no ivory towers. Scientific "objectivity" and "disinterest" become at best ignorance, at worst moral failing and spiritual disaster.

ISLAMIC SCIENCE

If this view of things is inherent to Islamic cosmology, why was Islamic science the most advanced in the world for several centuries? The very formulation of this question raises several issues that need to be considered before any attempt is made to answer (here Nasr's *Science and Civilization in Islam* can be consulted with profit).

First, the modern historians of Islamic science have believed implicitly if not explicitly in scientific progress, and they measure "advancement" in terms dictated by this belief. The earlier historians were interested in the texts mainly because of their "scientific" content, and they ignored everything that they considered theological, mystical, or superstitious – just as they discarded most of Newton's works so as to preserve his respectability as the father of modern science. Many historians have continued to study Islamic science with at least the partial aim of discovering why it did not follow the same enlightened route that science followed in the West, as if modern science is by definition normative and has brought about unquestioned benefit.

Second, even if we grant that some of the Islamic texts are "scientific" in a modern sense, their cultural context is every bit as important as their overt content. How did Ibn al-Haytham or al-Bīrūnī understand their own scientific works? Was their optics, mathematics, astronomy, and geology totally distinct from their metaphysics and spiritual psychology? And more importantly, how were their works read by their contemporaries? The work of the medieval

Muslim "scientists" was understood in terms of the dominant worldview of the time.

Third, the modern Western tradition has ascribed the highest value to rational thinking, but rationalism has in fact played a more restricted role in Islamic history than many historians suggest. Both Muslim apologists and Western scholars have highlighted the rational sciences in the Islamic past. Early Western scholars were busy tracing the origins of the types of thinking that they considered significant. They sought the causes of what they thought was aborted progress in the conflict between the "free thinking" of the philosophers and the "orthodoxy" of the theologians and jurists. On the Muslim side, the apologists have been eager to show that at the beginning, Muslims were enlightened, rational, good people, and then they were diverted from their glorious heights of scientific progress by sinister forces, if not foreign invasions. It was not Islam, they tell us, but the un-Islamic intrusions that led Muslims to abandon scientific progress and devote themselves to obfuscation and darkness.

If we look at the Qur'an and the way in which it has been interpreted by the Islamic community as a whole – not just by its rationalistically oriented theologians and jurists – we see that it stresses both God's utter transcendence and his total and intimate control of the universe. To speak of "control," however, is to use a scientific, rational term. We would do much better to speak of God's presence in all things through his signs, or the radiance of his infinite *wujūd*. The net result of understanding God as both absent and present led to the establishment of the two complementary modes of understanding that Ibn 'Arabī called "reason" and "imagination." Knowledge through rational processes stressed God's distance and transcendence. Knowledge through direct perception of God's presence in the things, or through the "symbolism" of things, stressed instead his nearness and immanence.

The rational approach seems almost "scientific," and it is this that has been the focus of studies for most Western scholars and the Muslim modernists/fundamentalists. The symbolist approach – branded "mystical," "irrational," and "superstitious" by the same people – came to be looked upon with contempt and was dismissed by Muslims as un-Islamic. If it is un-Islamic, true Islamic cosmology can be recovered by ridding Islamic thought of the vestiges of Qur'anic language and pushing God as far as possible from the universe. Then there will be no necessity to pay any heed to the soldiers of *wujūd*, and it will be easy to justify the technological rape of the earth and the electronic impoverishment of the human soul – so long as lip service is paid to the Qur'an, the Sunnah, and the Shariah.

THE EFFICACY OF NAMES

It was said earlier that names are efficacious by nature. Scientific names allow us to think of things "scientifically," which means that we can dismiss anything but quantifiable reality. Islamic reality is not quantifiable, which is to say that real things possess the attributes of life, knowledge, desire, power, speech, hearing, seeing, and so on, and the degree to which they possess them has nothing to do with "quantity" and everything to do with "quality." These attributes are simultaneously divine, cosmic, and human. Things make them manifest through a subtle and immeasurable participation in the radiance of the Real *wujūd*. Attributes that pertain to human beings also pertain to non-human things – including totally inanimate "objects" – because they pertain to God, the Creator of all things, "the Light of the heavens and the earth" (Qur'an 24:35), who sends down his light on everything in a measure known only to himself.

Once things have been named, we deal with them as their names allow. Cultural anthropology has illustrated the arbitrariness with which names can be given to things – especially if we take "rational" or scientific nomenclature as normative. But scientific nomenclature is itself arbitrary when viewed from the standpoint of any of the traditional cultural matrices, which bestow orientation on human beings by naming things in the context of grand master schemes of meaning. What appears arbitrary to Islamic thinking is any system of naming that ignores the transcendent dimensions to things and wrenches them from their qualitative contexts. It is these contexts that allow us to see how they are connected with greater wholes, with the world of the Unseen, and with the ultimately Real.

The governing insight of Islamic thinking, after the assertion of the unity and ultimacy of the Real, is that the true nature of the world is inaccessible to human beings without help. This insight is made explicit in the second half of the Shahadah, though it is also implicit in the first. Without messengers from the Real, no one can come to know God and the theomorphic roots of human nature. Indeed, it is not difficult to see that it is precisely the rejection of human dependence on the One that brought about the great split between the modern West and traditional religion. Take, for example, Toby E. Huff's summary of the metaphysics of modern science:

> We must keep in mind that the modern scientific worldview is a unique metaphysical structure. This means that the modern scientific worldview rests on certain assumptions about the regularity and lawfulness of the natural world and the presumption that man is capable of grasping this underlying structure. ... [M]odern science is a metaphysical system that asserts that man, unaided by spiritual agencies or divine guidance, is single-handedly capable of understanding and grasping the laws that govern man and the universe. The evolution of this worldview has long been in process, and ... we in the West simply take it for granted. ... The rise of

> modern science was not just the triumph of technical reasoning but an intellectual struggle over the constitution of the legitimating directive structures of the West.[15]

One of the primary "legitimating directive structures" of any culture is provided by the transmitted names of things. The breakthrough to modern science occurred when people learned how to name things on their own, without reference to the foundational myths of society, but this modified the efficacy of the naming. Having assumed full responsibility for naming, people remained blind and deaf to the Real and could not see beyond their own physical, social, and cultural horizons.

In Islamic terms, the fact that God names himself is the key to the extraordinary efficacy of the revealed names – their ability to chart a happy course not only through this life but also through all the worlds that follow death. God's primordial act of naming took place when he taught the names to Adam, and he has kept these names alive by sending 124,000 prophets down to Muhammad. It is as if, by naming the cosmos, he bestowed sight on the blind. As al-Ghazālī puts it, the Qur'an in relation to intelligence is like the sun in relation to the eye.

By naming the cosmic order, God allows people to see its significance in the whole of reality. By naming the human order, he allows people to see their proper role in society and nature. By naming human attributes, he allows people to grasp the difference between sick and healthy souls. By naming right and wrong, he allows morality and ethics to have an efficacy that transcends limited human views of the world and society. The overarching order in all these domains can never be grasped by strictly human means, because the overarching order is the Real itself, the ultimately unnamable and unknowable. Unless human beings acknowledge the names that the Real has bestowed, they will live in the darkness of misnomers.

INADEQUATE NAMES

From the standpoint of the intellectual tradition, the peculiar course of modern history is driven by the systematic application of inadequate names. No one doubts that such names have an efficacy all their own. The enormous power of modern technology and the unprecedented coerciveness of modern institutions became possible only when the human, anthropomorphic names were relegated to the domain of superstition and, at the same time, the "real names" were found through quantification and scientific analysis.

Quantification makes perfect sense in the context of mechanism, and conceiving of reality as a machine allows for manipulation without any restraints but the mechanical. It is no accident that ideology is commonly recognized as the blueprint for "social engineering." When things and people are looked upon as mere objects, reality is perceived as objective and impersonal, and this demands that we treat things with objectivity and disinterest. If the immediate is impersonal, so also must be the ultimate. In contrast, anthropomorphizing – especially as carried out by those who see themselves as theomorphic – diverts people from contemporary "reality" and prevents them from becoming docile production-line workers and hard-nosed doctors, engineers, and CEOs. Hence the real danger of "Sufism" for Muslim modernists and fundamentalists.

Once I heard Nasr say in a lecture – no doubt with a touch of Oriental hyperbole – that as soon as a Muslim schoolboy learns that water is H_2O, he stops saying his daily prayers. I offer my own commentary.

The traditional view of the cosmos presses upon people the interrelatedness of the divine, cosmic, and human orders. The daily prayers that God commands people to perform are nothing but the natural activities of all of God's creatures. As the Qur'an puts it, "Have you not seen that everything in the heavens and the earth glorifies God, and the birds spreading their wings? Each knows its daily prayer and its glorification" (24:41). Water is not a substance to be quantified but a quality to be appreciated at every level of created reality. "God's throne is upon the water" (11:7); "Of water We made every living thing" (21:30); "He sends down out of heaven water, and each dry streambed flows in its own measure" (13:17). Water is one of the four elements, which is to say that it is one of the four qualities or characteristics that allow us to speak of diverse tendencies in the visible realm. All visible things are made of these four elements, but the elements combine in differing proportions, thus helping to determine each thing's aggregate of attributes. Earth keeps things stable and low. Water allows for movement, flow, and the penetration of light. Air is permeable, subtle, and naturally clear. Fire is inherently luminous, changeable, and ascending.

Such notions are standard fare in texts on cosmology and permeate the thinking of traditional Muslims. People know intuitively the qualities associated with the four elements, foods, and natural phenomena. Scientific thinking condemns such knowledge to superstition, or at best, condescends to recognize a certain poetic sensitivity.

When science is taught in the West, it is typically taught by believers in a scientific orthodoxy who never question the objective truth of their beliefs. But in Islamic countries, where the traditional worldview still clings to life, science is often taught by converts, and they are much more fervent than born believers in denouncing the superstition of the old ways. They consider it their moral duty to guide the young to the one and only truth. This sort of fervor is not so obvious in the West, though it does appear in cases like the debates between "creationists" and "evolutionists," where the latter exhibit all the indignation of Puritan preachers – if the former do too, well, that is hardly remarkable.

The Islamist rhetoric that nowadays accompanies the teaching of science and other subjects is designed to wrench the remaining traditional teachings from their context and to politicize the students in keeping with current ideology. Such rhetoric simply hastens the reification of the cosmos by diverting Islamic sensibilities into an alien but very modern sphere. The mullah regimes have changed nothing here. They are just as enamored of the scientific worldview as anyone else, and in any case, the teachers are the same teachers. They have simply learned to toe the new party line, which now means spouting religious pieties, whereas before it had meant reciting political slogans. The official, government worldview, though labeled "Islamic," is now totally politicized, and it owes its genealogy to the same ancestors that have given us the ugliest forms of totalitarianism.

In short, the Muslim boy who is taught that water is really just H_2O learns that the qualities his grandmother sees in things and the names she applies to them are primitive and superstitious, and he jettisons her understanding along with all its accouterments, including her daily prayers. If, nowadays, Muslims boys have started to pray again, as likely as not they are acknowledging their allegiance to the Islamist party, or protecting themselves against the real dangers of political nonconformity in a coercive society.

THE MYTH OF THE UNSEEN MEN

The notion of the Men of the Unseen is a potent way of presenting basic themes of the Islamic worldview in a coherent myth and showing the inseparability of cosmos and soul. Let me illustrate by discussing four of these themes: unity, bilateralism, hierarchy, and theomorphism.

As we have seen, *tawḥīd* recognizes two modalities: first, transcendence and absoluteness, the fact that God is uniquely and utterly one and real; and second, immanence and infinity, the fact that everything is embraced by God's oneness and displays his attributes. The simultaneous oneness of the Real and manyness of creation are prefigured *in divinis* by the divine names, each of which designates the One along with a specific quality of the One, different from every other quality.

By speaking of the Men of the Unseen, Sufis assert God's transcendence and uniqueness by putting God at the pinnacle, beyond the universe, and they assert his immanence and polynomiality by conceiving of the basic structure of the universe in terms of human functions, each of which manifests various divine attributes. Within the created order, God's unity is reflected in the fact that the Pole is always one, and the hierarchy of God's names is reflected in the fact that the Men of Number are ranked in degrees below the Pole.

The mathematical progression of the Men – such as 1, 2, 4, 7, 12 – reflects the modes in which the divine Principle unfolds its potentialities through a hierarchy of created realities. Cosmically, these numbers can be discerned in the structure of natural phenomena throughout the universe. We have here a traditional mathematical scheme, but one that is hardly abstract, since anyone can grasp it immediately by reflecting on the world. The number one appears in the unity of each individual thing; two in day and night, heaven and earth, light and darkness; four in the elements, the seasons, the directions, the humors; seven in the planets; twelve in the zodiac.

Some authors explain the Unseen Men by illustrating the interrelationship of all things in terms of the divine names. Thus, for example, the Pole manifests the name God, because the Pole is the fully actualized image of God, comprehending and embodying all the divine attributes without exception. The two Imams manifest the names king and lord – that is, God as ruler and controller of the universe (the Absolute) and God as nurturer and protector of each thing in the universe (the Infinite). The four Pegs display the traces of the names alive, knowing, desiring, and powerful (sometimes called "the four pillars" of the divinity). The seven Substitutes reveal the properties of the names alive, knowing, loving, powerful, grateful, hearing, and seeing ("the seven leaders").

The bilateralism of transcendence and immanence is already implicit in the term "Men of the Unseen" because "unseen" is the conceptual counterpart of "visible," and the two together designate the two primary worlds. The visible world is the body of the cosmos, the unseen world its spirit. Like all bodily things, the visible world is indefinitely divisible, and its predominant characteristics are multiplicity, grossness, opacity, fragility, evanescence, change. In contrast, the unseen world partakes of unity, subtlety, luminosity, strength, permanence, fixity.

These specific attributes, however, are applied to the two worlds when they are envisaged in relatively impersonal terms. In fact, the unseen domain partakes of all the personal, divine attributes in a direct and active mode. Hence it is alive, knowing, desiring, powerful, speaking, hearing, seeing, merciful, forgiving, vengeful. These attributes can hardly be found in the visible world itself, though we are familiar with their traces. We notice them when we deduce the unseen attributes that motivate visible activities. Some acts suggest generosity, some vengefulness, some compassion. The fact that these attributes derive from the unseen realm is acknowledged by expressions such as "soul" and "spirit." The full actualization of these attributes can only be sought in the unseen, which helps explain the special characteristics of angels.

The many and diverse Sufi expositions of the nature of the cosmos are much more explicit than those of the philosophers in explaining the utterly central role of human beings for cosmic reality itself. Modern sensibilities dismiss such views for many reasons, not least because they seem to ignore the vast reaches of the universe brought to light by modern scientific techniques. But the Sufis were well aware that our specific world has no great significance in the overall scheme of things and that the universe is unlimited in time or space, except inasmuch as its createdness differentiates it from the Uncreated, which is "infinite" in the strict sense.

What appears truly strange to Islamic cosmology is that the scientific universe is considered to be all that there is, when in fact it can only be an infinitesimal speck in all of reality (as Hindu and Buddhist cosmologies know so well). The scientific universe is "physical" by definition, which is to say that it is simply what is available to the science of physics, which can never go beyond a "sensualist and empirical epistemology." The metaphysical and methodological presuppositions of physics specifically and science generally allow for no access to the invisible realm of pure intelligence, the intense radiance of self-aware *wujūd*. In other words, science can provide no direct insight into the nature of the unseen realm that is the home of the human spirit.

In the Islamic worldview, the relationship between the Unseen and the Visible is analogous to that between God and the cosmos. The Unseen is infinitely more vast, powerful, active, intelligent, conscious, and compassionate than the Visible, even though the two worlds together are as nothing compared to God. Since human beings have the peculiar characteristic of being made in God's image, they are also images of his whole creation, which is the sum total of the unseen and visible realms. Just as the unseen realm of the cosmos is far more real than its visible realm, so also the unseen realm of human beings is far more real than their visible realm.

We recognize the superior reality of our own unseen dimension precisely to the extent that we find human significance in qualities such as love, compassion, wisdom, understanding, forgiveness, generosity, discernment, justice, and pardon – qualities not found per se in the visible world, but nonetheless traditionally understood as essential characteristics of the divine and the human. To say that these unseen qualities pertain to a higher order of reality means that the more intensely these qualities are found, the more intensely reality is present. In no way are they "epiphenomena" of the human order or any other order. To look at the universe in that way is to invert the normal and normative order of things; it is to take the highest as the lowest, and the lowest as the highest.

Because human beings are made in the divine image, they have the potential to manifest all the divine names in diverse degrees of intensity. They differ radically from all other creatures by possessing a synthetic and all-comprehensive nature, which allows them to manifest the most fundamental divine qualities in a fullness that is inconceivable in any other mode of being, unseen or visible. Compassion, love, justice, and forgiveness are qualities actualized in the human image of the divine, and they are not found anywhere else in the universe as we know it, except in dim and metaphorical modes. It follows that human beings are the most real beings in the cosmos that we know. What modern scientism would call "objective reality" is as impermanent, evanescent, and insignificant as a cloud – as many physicists have been telling us.

The only permanent reality, the only thing that is truly real, is the Real itself. Its attributes become manifest to significant degrees only in the unseen domain, the realm of consciousness, awareness, life, love, compassion, justice. What appears as "epiphenomena" to the proponents of scientism is the face of reality itself, hidden behind the veil of phenomena, and what appears as real is a fading illusion.

Where is the "real world"? Only in the Unseen, and it is fully actualized only in the unseen realm of human beings. Even angels, though they dwell in the Unseen, are peripheral beings, which explains why God commanded them to prostrate themselves before Adam after he had taught him the names. Human beings alone can name reality in its fullness, because their inmost nature has access to every name God has taught. When they name things as the Real names them, they necessarily name the unseen realm as primary and most significant. This explains why those among them who have traditionally been recognized as the wisest and most humane have consistently affirmed the overriding reality of the Unseen – the hidden, divine attributes that need to be made manifest in terms of visible, social reality, through compassion, love, morality, ethics, and law.

The Unseen Men do not live in the visible world. They live with God, who manifests himself most directly in the unseen realm. Just as human beings play a central role in the visible realm, effectively ruling over the world by taking an active role vis-à-vis the relative passivity of all other creatures, so also they play a central role in the unseen realm, since the great ones among them rule the world of consciousness and awareness. The grand difference between the two types of rulership is that in the visible realm, rulership too often follows the whims of individuals and the vagaries of human institutions. In the unseen realm, human rulers follow the divine King in perfect harmony. Those who deny or reject the authority of God's self-naming, or those who misinterpret it for their own aims, may attempt to govern the visible world according to their own misnomers. But those who name things with God-given names deal with them exactly as God himself is dealing with them through his continual and ongoing creation of the universe. True control belongs to God alone, no matter who appears to be in charge. The Men of the Unseen acknowledge their utter submission to him and play the role of his vicegerents in governing the invisible affairs that control the visible realm.

In this scheme of things, problems arise only from human misunderstanding or misapplication of the divine names (Satan also plays a role, but not without human intermediary). The Muslim view allows for no despair, however, because it recognizes that God's mercy takes precedence over his wrath, and that, in the last analysis, he holds the universe in mercy's hand. Those who fail to follow his instructions by submitting to him voluntarily, but who instead, like Satan, embark on their own courses, fit nonetheless into the divine scheme of things, and in the end, God's wisdom will be perceived in even the worst of men and the worst of evils. Everything will be well, but not according to our lights – unless, of course, our lights have submitted to the divine Light.

It is the recognition of this underlying mercy and compassion, I think, that allows Nasr always to put the best spin on things. Those who know him personally know that he always sees the good side of people and events – contrary to what might be expected from the critical tone of some of his writings. Certainly, he never suggests that anyone should stop trusting in God's wisdom and compassion. At the same time, he asks people to take advantage of the best in themselves in order to rethink their relationship with God and the world. On this note, I will let Nasr have the last word. In one of his recent books he offers the Unseen Men's solution to the impasse that modern humanity has constructed for itself:

> What is needed is a rediscovery of nature as sacred reality and the rebirth of man as the guardian of the sacred, which implies the death of the image of man and nature that has given birth to modernism and its subsequent developments. It does not mean the "invention of a new man" as some have claimed, but rather the resurfacing of the true man, the pontifical man whose reality we still bear within ourselves. Nor does it mean the invention of a sacred view of nature, as if man could ever invent the sacred, but rather the reformulation of the traditional cosmologies and views of nature held by various religions throughout history. It means most of all taking seriously the religious understanding of the order of nature as knowledge corresponding to a vital aspect of cosmic reality and not only subjective conjectures or historical

constructs. There must be a radical restructuring of the intellectual landscape to enable us to take this type of knowledge of nature seriously, which means to accept the findings of modern science only within the confines of the limitations that its philosophical suppositions, epistemologies, and historical development have imposed upon it, while rejecting completely its totalitarian claims as *the* science of the natural order. It means to rediscover a science of nature that deals with the *existence* of natural objects in their relation to Being, with their subtle as well as gross aspects, with their interrelatedness to the rest of the cosmos and to us, with their symbolic significance and with their nexus to higher levels of existence leading to the Divine Origin of all things.[16]

6

The Anthropocosmic Vision

I take the expression "anthropocosmic vision" from Tu Weiming, Professor of Chinese History and Philosophy and Confucian Studies at Harvard University, and Director of the Harvard-Yenching Institute. Professor Tu has used it for many years to encapsulate the East Asian worldview and stress its salient differences with the theocentric and anthropocentric worldviews of the West.[17] By saying that the Chinese traditions in general and Confucianism in particular see things "anthropocosmically," he wants to say that Chinese thinkers and sages have understood human beings and the cosmos as a single, organismic whole. The goal of human life is to harmonize oneself with heaven and earth and to return to the transcendent source of both humans and the world.

As long as Chinese civilization retained its anthropocosmic vision, it could not develop instrumental rationality, the Enlightenment view that sees the world as a conglomeration of objects and understands knowledge as the means to control the world. In the anthropocosmic vision, the object cannot be disjoined from the subject. The purpose of knowledge is not to manipulate the world but to understand the world and ourselves so that we can live up to the fullness of our humanity. The aim, to use one of Tu Weiming's favorite phrases, is "to learn how to be human." As he writes, "The Way is nothing other than the actualization of true human nature."[18]

With slight revisions in terminology, Tu Weiming's depiction of the anthropocosmic vision could easily be employed to describe the overarching worldview of Islamic civilization in general and the intellectual tradition in particular. For the purposes of this chapter, I will focus more on the philosophical side of the tradition. I do so because, first, among all the Islamic approaches to knowledge, philosophy has produced figures who have been looked back upon by Western historians and modern-day Muslims as "scientists" in something like the current meaning of the word; and second, only this approach has discussed the significance of being and becoming without presupposing faith in Islamic dogma, so its language can more easily be understood outside the context of specifically Islamic imagery.

AHISTORICAL AND HISTORICAL KNOWLEDGE

In Western civilization, a sharp distinction has commonly been drawn between reason and revelation, or Athens and Jerusalem. In order to understand the role that the intellectual sciences have played in the Islamic tradition, we need to understand that the predominant Islamic perspective has seen reason and revelation as harmonious and complementary, not antagonistic. The very content of the Qur'anic message led to a viewpoint that diverges sharply from what became normative in the Christian West. Without understanding the divergent viewpoint, we will find it difficult to grasp the role that wisdom has played in Islam.

If we look at Christianity in terms of the dichotomy between intellectual and transmitted knowledge, what immediately strikes the eye is that the fundamental truths are indebted to transmission, not intellection. The defining notion of the Christian worldview – to the extent that it is meaningful to generalize about a complex and many-sided tradition – is the incarnation, an historical event whose occurrence is known through transmitted knowledge. To be sure, the incarnation was seen as a divine intervention that transmuted history, but it was also understood as occurring in the full light of historical actuality. In order to know about it, people needed the transmission of historical reports.

The Islamic tradition has a very different starting point. It is often assumed by both Muslims and non-Muslims that Islam began with the historical event of Muhammad and the Qur'an. There is some truth in this, of course, but the Qur'an paints a different picture, one that has had a deep effect on the way people have conceived of their religion. In this perspective, Islam began with the creation of the world. In its broadest meaning, the word *islām* (submission, submittedness, surrender) designates the universal and ever-present situation of creatures in face of the Creator. "To Him is submitted everything in the heavens and the earth" (Qur'an 3:83). This helps explain why the first and fundamental dogma of the religion is *tawḥīd*, which has nothing to do with the historical facts of Muhammad and the Qur'an.

Tawḥīd is the acknowledgment of a universal truth that expresses the actual situation of all things for all time and all eternity, since everything submits to God's Unity by the very fact of its existence. Only human beings among all creatures have the peculiar status of being able, in a certain respect, to accept or reject this truth. To accept it freely is to utter the first half of the Shahadah and give witness to the unique reality of God. The Qur'an attributes *tawḥīd* and the free acceptance of its consequences to all rightly guided people, beginning with Adam and extending down through all the prophets and all those who correctly and sincerely followed them.

It might be objected that the statement of *tawḥīd* is itself historically particular. But the issue is not its linguistic formulation, but rather the unique, unitary reality that gives rise to the universe. Note that the Qur'an says that God sends every message in the language of the prophet's people (14:4) and that "Each community has a messenger" (10:47). The basic content of every message was *tawḥīd*: "And We never sent a messenger before you save that We revealed to him, saying, 'There is no god but I, so serve Me' " (21:25). "There is no god but I" is the first truth of every message, the first half of its Shahadah. "Serve Me" lays down the necessity for a second Shahadah to delineate the specific forms of "service" (*'ibāda* = "worship") appropriate to the cultural and historical context of the people to whom the message is addressed.

One might also object that this unitary reality is itself historically particular, because it was invented by human minds. People who hold this position still have to justify it, so they cannot escape a metaphysics. On what basis do we declare history, language, politics, gender, atoms, energy, the brain, genes, or whatever else foundational? Notice, moreover, that such theories are always rooted in forms of transmitted knowledge that go back to historical authorities who function as prophets for believers in the theories. One is reminded of the old joke, heard among scholars of Islam at least, that Marxism boils down to this Shahadah: "There is no god, and Karl Marx is his messenger."

In the Islamic perspective, *tawḥīd* stands outside history and outside transmission. It is a universal truth that does not depend on revelation. Understanding it is an inherent quality of the innate disposition (*fiṭra*) of Adam and his children. The fall from paradise does not represent a serious shortcoming, but rather a temporary lapse, a single act of forgetfulness and disobedience. The lapse had repercussions to be sure, but it was immediately forgiven by God, and Adam was designated as the first prophet. His divine image was in no way blemished by the fall, even if it does become obscured in many if not most of his children.

Tawḥīd precedes Muhammad and his revealed message, because it does not pertain to history. It informs all true knowledge in all times and all places. Every one of the 124,000 prophets brought it as the basis of the message. They did not teach it, however, in order to establish an authoritative belief system that could be transmitted to others. Rather, they taught it because people have a tendency to forget it and need to be "reminded" (*dhikr*).

This word *dhikr* (along with its derivatives *tadhkīr*, *tadhkira*, and *dhikrā*) designates one of the most important concepts in the Qur'an. It informs Islamic religiosity on every level of faith and practice. It means not only "to remind," but also "to remember." In the sense of reminder, it indicates the primary function of the prophets, and in the sense of remembrance it designates the proper human response to the prophetic reminder. The whole process of learning how to be human depends first upon being reminded of *tawḥīd*, and second upon active and free remembrance.

If the first half of the Shahadah stands outside history, the second half – "Muhammad is God's messenger" – is firmly grounded within it. It refers to the historical particularities of the Islamic tradition, which began in the seventh Christian century with the revelation of the Qur'an. Thus the two halves of the Shahadah implicitly distinguish between a universal, ahistorical truth and a particular, historical, and conditioned truth. Simultaneously, they distinguish between intellectual and transmitted knowledge. The first half articulates a knowledge innate to the original human disposition and accessible to all human beings; and the second half establishes the authority of a specific, historical message embodied in the Qur'an, the message of Muhammad, with all its detailed teachings.

THE PHILOSOPHICAL QUEST

Among all the schools of Islamic thought, the philosophers were the most careful to distinguish between transmitted and intellectual learning. They themselves were not primarily interested in transmitted knowledge. Compared to jurists, theologians, and Sufis, philosophers paid little attention to the Qur'an, the Hadith, and the religious sciences. It is true, nonetheless, that most of them were well versed in the transmitted religious learning, and some even wrote Qur'an commentaries and juridical works. They were not hostile to the transmitted learning, but rather focused their attention elsewhere. They wanted to develop their own intellectual vision by working out the implications of *tawḥīd* in theory and in practice.

The philosophers undertook the quest for wisdom with the ultimate aim of transforming their souls. As Tu Weiming says of the Confucian anthropocosmic vision, "The transformative act is predicated on a transcendent vision that ontologically we are infinitely better and therefore more worthy than we actually are."[19] This is a "humanistic" vision, but a humanism that is elevated far beyond the mundane, because the measure of all things is not man or even rational understanding, but the transcendent source of all. As Tu puts it,

> Since the value of the human is not anthropocentric, the assertion that man is the measure of all things is not humanistic enough. To fully express our humanity, we must engage in a dialogue with Heaven because human nature, as conferred by Heaven, realizes its nature not by departing from its source but by returning to it. Humanity, so conceived, is the public property of the cosmos, not

> the private possession of the anthropological world, and is as much the defining characteristic of our being as the self-conscious manifestation of Heaven. Humanity is Heaven's form of self-disclosure, self-expression, and self-realization. If we fail to live up to our humanity, we fail cosmologically in our mission as co-creator of Heaven and Earth and morally in our duty as fellow participants in the great cosmic transformation.[20]

For the Islamic wisdom tradition, grasping the full nature of our humanity necessitates investigating the nature of things and the reality of our own selves. This meant that intellectuals could not limit themselves to the mere acceptance of transmitted learning. They could not ignore the human imperative to search for knowledge in every domain, especially not when the Qur'an explicitly commands the study of the cosmos and the soul as the means to know God. Although some philosophers paid little attention to the transmitted learning and had no patience with the quibbling of theologians and jurists, they did not step outside of Islam, because they could not doubt the universal and ahistorical axiom upon which it is built. In other words, there was no historical chink in their intellectual armor. Historical contingencies cannot touch *tawḥīd*, because, once it is grasped, it is seen as so foundational that it becomes the unique certainty upon which the soul can depend.

As for the theologians and jurists and their claims to authority in religious matters, the representatives of the wisdom tradition saw those claims as pertaining to transmitted learning, not intellectual learning, and they found no reason to submit themselves to the limited understandings of pious dogmatists. To a large degree they kept themselves apart from theological and juridical bickering, and this helps explain why the philosophers among them (in contrast to the Sufis) preferred to employ a language colored more by Greek models than the imagery and symbols of the Qur'an.

Once we recognize that Islamic intellectual learning stands aloof from transmitted learning, it becomes clear why the modern scientific enterprise could not have arisen in Islam. Science gains its power from rejection of any sort of teleology, brute separation of subject and object, refusal to admit that consciousness and awareness are more real than material facts, exclusive concern with the domain of the senses, and disregard for the ultimate and the transcendent. Instrumental rationality could appear in the West only after the baby had been thrown out with the bath water. Having rejected the bath water of theology – or at least the relevance of theological dogma to scientific concerns – Western philosophers and scientists also rejected the truth of *tawḥīd*, the bedrock of human intelligence. Once *tawḥīd* was a dead letter, every domain of learning could be considered an independent realm.

Instrumental rationality did not appear suddenly in the West, of course. A long and complex history gradually brought about an increasingly wider separation between the domains of reason and revelation. Many scientists and philosophers remained practicing Christians, but this did not prevent them from considering the rational domain free from the trammels of revealed givens. It is precisely because these givens were posed in the dogmatic and historical terms of transmitted learning rather than the open-ended and ahistorical terms of intellectual learning that the separation between reason and revelation could occur.

In contrast, the Muslim intellectuals kept themselves rooted in the vision of *tawḥīd*. No matter what sort of misgivings some of them may have entertained concerning the historical contingency of the Arabic language, the events surrounding the appearance of Muhammad, the transmission of the Qur'anic revelation, and the interpretation of the revelation by the theologians and dogmatists, they did not see these as impinging on the fundamental insight of *tawḥīd*, which for them was utterly transparent.

My first conclusion, then, is this: many historians have suggested that medieval Islamic learning declined when Muslim scientists neglected to build on their early discoveries. But this is to read Islamic history in terms of the ideology of progress, which in turn is rooted in contemporary scientism – the belief that science has the same sort of unique reliability that was once reserved for revealed truth. Scientism gives absolute importance to scientific theories and relativizes all other approaches to knowledge. This is not to deny that there was a decline in Islamic learning; it is simply to call into question the criteria by which such things are normally judged. Why should historical oddities such as the ideological presuppositions of modernity be the yardstick for civilization? If we keep in view Islamic criteria (e.g., adherence to *tawḥīd*, the Qur'an, and the Sunnah), there was certainly a serious decline, but that decline cannot be measured by the criteria that are normally applied.

Moreover, historians who talk in broad terms of the decline of Islamic "science" fail to acknowledge the profound difference between two historical contexts. The first is the Islamic, in which the axiom of *tawḥīd* infused all intellectual endeavor. The philosophers saw all things as beginning, flourishing, and ending within the compass of the One Source, so they could not split up the domains of reality in more than a tentative way. They were not able to disengage knowledge of the cosmos from knowledge of God or knowledge of the soul. It was impossible for them to imagine the world and the self as separate from each other or from the One Principle. Quite the contrary, the more they investigated the universe, the more they saw it as displaying *tawḥīd* and the nature of the self. They could not have agreed more with Tu Weiming, who writes, "To see nature as an external object out there is to create an artificial barrier which obstructs our true vision and undermines our human capacity to experience nature from within."[21]

The second context that people tend to forget when they claim that the Muslim intellectual tradition declined is the Christian. Christian civilization, qua *Christian* civilization, did in fact decline and, many have argued, disappeared, because it experienced the breakdown of a synthetic worldview. Part of the reason for this breakdown and the concurrent rise of a secular and scientistic worldview was that the transmitted nature of the basic religious givens was not able to withstand the critical questioning of non-dogmatic thinkers. In the Islamic case, Muslim intellectuals did not depend on revelation and transmission for their understanding of *tawḥīd*, so theological squabbles and historical uncertainties could not touch their basic vision of reality.

THE METHODOLOGY OF *TAḤQĪQ*

In order to suggest some of the implications of the anthropocosmic vision, I need to expand a bit more on the distinction between intellectual and transmitted. The experts in transmitted learning claimed authority for their knowledge by upholding the truthfulness of those who provided the knowledge – that is, God, Muhammad, and the pious forebears – and the authenticity of the transmission. They asked all Muslims to accept this knowledge as it was received. The basic duty of the Muslim believer was *taqlīd*, imitation or submission to the authority of the transmitted knowledge. In contrast, the intellectual tradition appealed to the relatively small number of people who had the appropriate aptitudes. The quest for knowledge was defined in terms of *taḥqīq*, verifying and realizing the truth for oneself.

If we fail to see that knowledge achieved by realization is not of the same sort as knowledge received by imitation, we will not be able to understand what the Muslim intellectuals were trying to do or what modern scientists and scholars are trying to do. We will continue to falsify the position of the Muslim philosophers by making them precursors of modern science, as if they were trying to discover what modern scientists try to discover, and as if they accepted the findings of their predecessors on the basis of imitation, as modern scientists do.

Given that scientism infuses modern culture, it is difficult for moderns to remember that the whole scientific edifice is built on transmitted learning. Despite all the talk of the "empirical verification" of scientific findings, this verification depends on assumptions about the nature of reality that cannot be verified by empirical methods. Even if we accept for a moment the scientistic proposition that scientific knowledge is uniquely "objective," it is in fact verifiable only by a handful of specialists, since the rest of the human race does not have the necessary training. In effect, everyone has to accept empirical verification on the basis of hearsay. As Appleyard puts it, "Scientists who insist that they are telling us how the world incontrovertibly is are asking for our faith in their subjective certainty of their own objectivity."[22]

It was noted that the word *taḥqīq* derives from the word *ḥaqq*, meaning true, truth, real, right, proper, just, appropriate. When the word *ḥaqq* is applied to God, it means that God is the absolutely true, right, real, and proper. But the word is also applied to everything other than God. This secondary application acknowledges that everything in the universe has a truth, a rightness, a realness, and an appropriateness. God is *ḥaqq* in the absolute sense, and everything other than God is *ḥaqq* in a relative sense. The task of *taḥqīq* is to build on the knowledge of the absolute *ḥaqq*, beginning with the axiom of *tawḥīd*, and to grasp the exact nature of the relative *ḥaqq* that pertains to each thing, or at least to each thing with which we come into contact, whether spiritually, intellectually, psychologically, physically, or socially.

The formula of *tawḥīd* tells us that there is no god but God, no *ḥaqq* but the absolute *ḥaqq*. This *ḥaqq* is transcendent, infinite, and eternal, and nothing else can be worthy of the name. Nonetheless, all things are creations of God, having received everything that they are from him. He creates them with wisdom and purpose, and each has a role to play in the universe. Nothing that exists is inherently *bāṭil* – false, vain, unreal, inappropriate. This is not to say that there is no such thing as "evil." The issue of discerning the *ḥaqq* of "evil" is one of the more subtle dimensions of *taḥqīq*. Recognizing a thing's *ḥaqq* may well entail acknowledging that part of its proper role is to be an occasion for evil and that the appropriate human response is to avoid it. The very need to avoid evil alerts us to something of its cosmic function: its possibility bestows meaning and significance on human freedom.

The *ḥaqq* s of individual things are determined by God's wisdom in creation. It is in respect to these individual *ḥaqq* s that the Prophet said, "Give to each that has a *ḥaqq* its *ḥaqq*," a command that sums up the goal of *taḥqīq*. To achieve this is obviously more than a simple cognitive activity. We cannot give things their rightful due simply by knowing their truth and reality. Over and above knowing, *taḥqīq* demands acting. It is not simply to verify the truth and reality of a thing, it is also to act toward the thing in the appropriate and rightful manner.

Seekers of wisdom, then, were trying to verify and realize things. They could not do this by quoting the opinions of Aristotle or Plato, nor by citing the words of the Qur'an and Muhammad. They could take the prophets and the great philosophers as guides on the path to realization, but they could not claim to know what the prophets and sages knew unless they discovered it for themselves. The quest demanded training the mind and disciplining the soul. It demanded the achievement of an authentic vision of reality, a correct perception of the world, a sound understanding of the self, a true knowledge of the First Principle, and activity in terms of what one had come to know.

UNDERSTANDING THE SOUL

The intellectual tradition refers to the underlying substance of a human being as *nafs*, the basic reflexive pronoun in the Arabic language. The word is translated as "self" or "soul," depending on context. In its philosophical sense, it designates the invisible something that makes its appearance in the cosmos wherever there is life, and hence it is ascribed to every living thing.

Verifying the nature of soul was one of the foundational activities of the Muslim intellectuals. A standard way to do so was to begin by investigating the apparitions of soul in the visible world. The visible realm is a conglomeration of bodily appearances, yet we constantly differentiate among them in terms of their modalities of appearance. We know the difference between living things and dead things precisely by the way they appear to us. "Soul" is the generic name for what shows itself when we recognize life and awareness. When we recognize these qualities in things, we simultaneously recognize them in ourselves. It is soul that knows soul. We know a living thing because we are alive, and we recognize a self-acting thing because we have self-activity. What we see outside we find inside. Finding the external apparitions of soul is to experience the soul's presence to itself. Life and awareness are precisely properties that we find in ourselves in the very act of discerning them in others.

There are degrees of soul, which is to say that this invisible something is more intense and influential in some things than in others. As Tu Weiming writes about the Chinese understanding, "Rocks, trees, animals, humans, and gods represent different levels of spirituality based on the varying compositions of *ch'i*."[23] In the typical Islamic version, the *ch'i* or invisible power that animates rocks is called "nature" (*ṭabī'a*). Only at the plant level is a second modality of *ch'i*, "soul," added to the first. Nor are rocks "only matter." In the hylomorphism adopted by the intellectual tradition, the role of matter (*mādda*) is largely conceptual, because there is no such thing as matter per se. The name is given to an observed receptivity for the apparition of "form" (*ṣūra*). Form itself is an intelligible reality that descends into the realm of appearances from the spirit or intellect and ultimately from God, who is, in Qur'anic language, "the Form-giver" (*al-muṣawwir*). Since all things are "forms," there is nothing in the universe that does not manifest the living presence of intelligence and the intelligible.

The classification of creatures into inanimate, plant, animal, human, and angel is one way of acknowledging different degrees of soul. The most complex and layered soul is found in human beings. Outwardly, this appears in the indefinite diversity of their activities, which clearly has something to do with vast differences in aptitude and ability. Because of the diverse and comprehensive powers of human souls, people can grasp and replicate all the activities that other modalities of soul cause to appear in the world.

In discussing the human soul, the texts frequently elaborate on the intimate correspondence between soul and cosmos, which were understood in something like a subject–object relationship. The human soul is an aware subject that can take as its object the whole universe. So closely intertwined are soul and cosmos that, in Tu Weiming's term, their relationship can properly be called "organismic." They can be understood as one organism with two faces.

It follows that there can be no microcosm without macrocosm, and no macrocosm without microcosm. The vital cosmic role of human beings was always affirmed. It was recognized that the macrocosm appears before human beings, but it was also understood that the macrocosm is brought into existence precisely to make it possible for human beings to appear and for them to learn how to be human. Without human beings (or, one can surmise, analogous beings), there is no reason for a universe to exist in the first place. The teleology was always acknowledged.

In the more religious language, this is to say that God created the world with the specific aim of crowning his achievement with human beings, who alone are made fully in his image and are able to function as his vicegerents. They alone can love God, because they alone are able to embody every divine attribute. Genuine love demands loving the Beloved for himself, not for something less than he. If one loves God with the aim of receiving some gift or benefit, such as avoiding hell and going to paradise, one has not in fact loved God, but the gift or benefit. This may sound like a "Sufi" idea, but notice what Avicenna, the greatest of the Peripatetic philosophers, has to say about those who have entered the path of achieving self-knowledge:

> Knowers desire the Real, the First, only for His sake, not for the sake of something else. They prefer nothing to true knowledge of Him. Their service [*'ibāda*] is directed only to Him, because He is worthy of service, and service is a noble relationship with Him. At the same time, knowers have neither desire nor fear. Were they to have it, the object of desire or fear would be their motive, and it would be their goal. Then the Real would not be their goal but rather the means to something less than the Real, which would be their goal and object.[24]

In short, the only creature that can love God for God's sake alone, without any ulterior motive, is that which is made in his image. God created human beings precisely so that they can verify and realize their own divine images and love their Creator, thereby participating in his infinite and never-ending bounty.

For the intellectual tradition, the purpose of studying the macrocosm is to come to understand the powers and capacities of the microcosm. By understanding the object, we simultaneously grasp the potentialities and abilities of the subject. We cannot study the natural world without learning about ourselves, and we cannot learn about ourselves without coming to understand the wisdom inherent in the natural world.

Social reality was often studied for the same purpose – as an aid to understanding the human soul. It was not uncommon for Muslim philosophers to provide descriptions of the ideal society. But they were not interested in the utopian dreams which have so often preoccupied modern political theorists and which form the backbone of ideology. Rather, they wanted to understand and describe the various potentialities of the human soul that became manifest through social and political activity. They did not want to set down a program, but rather to illustrate to aspiring philosophers that every attribute and power of the soul, every beautiful and ugly character trait, can be recognized in the diversity of human types. When seekers of wisdom recognize their own selves as microcosms of society, they can strive to know and realize the true sovereign of the soul, the real philosopher-king, which is the intellect, whose duty is to govern both soul and body with wisdom and compassion.

If the philosophers analyzed the souls of plants, animals, humans, and even angels, and if they described all the possibilities of human becoming in ethical and social terms, their purpose was to integrate everything into the grand, hierarchical vision of *tawḥīd*. It was self-evident to them that the intellect within us – the intelligent and intelligible light of the soul – is the highest and most comprehensive dimension of the human substance. The intellect alone can see, understand, verify, and realize. The intellect alone gives life, awareness, and understanding not only to our own souls, but to all souls. The intellect alone is able to grasp and realize the purpose of human life and all life.

ORIGIN AND RETURN

What then is this intellect that is the fountainhead and goal of intellectual learning? To define it is impossible, because intellect is the very understanding that allows for definitions. It cannot be limited and confined by its own radiance. However, we can describe it in terms of its role in cosmogenesis, whereby all things are created through it. And we can also depict it in terms of the human return to God, which can be experienced in its fullness only by the actualized intellect, which is the self-aware image of God. Let me deal with cosmogenesis first.

The wisdom tradition typically discussed the birth of the cosmos as beginning with God's creation or emanation of the first creature, which is called by names like intellect, spirit, word, pen, and light. Things appear from the One Principle in a definite, intelligible order and in keeping with a fixed and known hierarchy (known, that is, to God and to the intellect, but not necessarily to us). It was obvious to Muslim thinkers that the One God creates intelligently, and that the first manifestation of his reality, the contingent being closest to his unity, the stage of created actuality nearest to his utter and absolute simplicity, is pure intelligence and awareness. Within this awareness are prefigured the universe and the human soul.

This living intelligence is the instrument through which the Real ordered, arranged, and established all creatures, and it liesat the root of every subject and every object. It is a single reality that is the self-aware and self-conscious source of the cosmos and the soul. Among all creatures, humans alone manifest its full and pure light, a light that the Qur'an calls the spirit that God blew into Adam. Inasmuch as the "fall" of Adam has a negative significance, it is nothing but the obscuration of this light.

When we look at the intellect from the point of view of the return to God, we see that the goal of human existence is to remember God by recollecting the divine image within the self and awakening the intellect. The task of seekers is to recover in themselves the luminous consciousness that fills the universe. This recovery is the fruition and fulfillment of human possibility. Although the intellect is dimly present in every soul, human or otherwise, in human beings alone is it a seed that can sprout and be cultivated, nourished, strengthened, and fully actualized.

The human soul is a knowing and aware subject that has the capacity to take as its object the whole cosmos and everything within it. However, it is typically blind to its own possibilities, and it takes on the color of souls that are not fully human. The soul needs to learn how to be human, and truly human activity does not come easy. Most of us have to be reminded about what being human implies, and even budding "intellectuals," with all their gifts, have a steep and rocky road ahead if they are to achieve the goal.

Part of learning how to be human involves differentiating the qualities of the human soul from the qualities of other souls, which represent limiting and confining possibilities of soulish existence. The moral injunctions to overcome animal instincts rise up from the understanding that animals cannot manifest the fullness of intellectual and ontological possibility. This is not to denigrate animal qualities, since they play positive and necessary roles in the world and in the human make-up. The issue is rather one of priorities. People need to put things in their proper places. They must order the world and their own selves in an intelligent manner, and this means that they must understand everything in terms of the ruling truths of the cosmos. They must give to everything that has a *ḥaqq* its *ḥaqq*, and all things have their *ḥaqq* s, both outside and inside the soul.

The soul, then, is the subjective pole of manifest reality, and its counterpart is the cosmos, the objective pole. The soul in its human form has the unique capacity to know all things. However, the soul possesses only the *potential* to know all things, not the actuality of knowing. Actuality is a quality of intellect. Every act of knowing actualizes the soul's potential and brings it closer to the intelligent and intelligible light at its core. But what exactly is the limit of the soul's potential? What can it know? What should it strive to know? The intellectual tradition answers that there is no limit to the soul's potential, because nothing exists that the soul cannot know. The goal of learning is to know everything that can possibly be known. However, knowable things need to be prioritized. If we do not search for understanding in the right manner and the correct order, the goal will remain forever unattainable. If we do not give knowing its *ḥaqq*, we will remain forever ignorant.

As long as the soul remains occupied with the search for wisdom and has not yet actualized its full potential, it remains a soul – that is, an aware self with the possibility of achieving greater awareness. Only when it reaches the actuality of all-knowingness in the innner-most core of its being can it be called an intellect in the proper sense of the word. At this point it comes to know itself as it was meant to be. It recovers its true nature, and it returns to its proper place in the cosmic hierarchy. The philosophers frequently call the human soul a "potential intellect" (*'aql bi'l-quwwa*) or a "hylic intellect" (*'aql hayūlānī*), which is to say that it has the capacity to know all things. Only after it has ascended through the stages of actualizing its own awareness and achieving its own innate perfection is it called an actual intellect.

Philosophers sometimes refer to the actualization of the intellect by employing the Qur'anic terms "salvation" (*najāt*) or "felicity" (*sa'āda*). They would agree with Tu Weiming, who writes, "Salvation means the full realization of the anthropocosmic reality inherent in our human nature."[25] For them, this anthropocosmic reality is the intellect that gave birth to both macrocosm and microcosm and that is innate to the human *fiṭra.*

OMNISCIENCE

If the Muslim philosophers saw the quest for wisdom as a search to know all things, can we conclude that they were simply imitating Aristotle, who says as much at the beginning of the *Metaphysics*? I think not. They would say that they are trying to live up to the human potential, and if Aristotle also understood the human potential, that is precisely why they call him "The First Teacher." They would remind us that the Qur'an discusses human potential in rather explicit terms. It tells us, after all, that God taught Adam *all* the names, not just some of them. They might also point out that this quest for omniscience is implicitly if not explicitly acknowledged not only by all the world's wisdom traditions, but also by the whole enterprise of modern science. But, from their perspective, omniscience can only be found in the omniscient, and the only created thing that is omniscient in any real sense is the fully actualized intellect, the radiance of God's own Selfhood. Omniscience, in other words, can never be found in the compilation of data, the collections of facts, and the spinning of theories. It is not an "objective" reality, but a "subjective" awakening – though no distinction can be drawn between subject and object when one has actualized the very being of the omniscient.

Nothing differentiates the Islamic intellectual quest from modern scientific and scholarly goals more clearly than the differing interpretations of the quest for omniscience. Both the Muslim intellectuals and modern scientists are striving to know everything, but the Muslim intellectual does so by looking at roots, principles, and noumena and by striving to synthesize all knowledge and to unify the knowing subject with its object. In contrast the modern scientist looks at branches, applications, and phenomena and strives to analyze objects, multiply data, and spin out theories.

The traditional intellectual undertakes the quest for omniscience as an individual. He knows that he must accomplish the task within himself and that he can do so only by achieving the fullness of humanity, with everything that this demands ethically and morally. The modern scientist undertakes his quest for facts and information as a collective undertaking, knowing that he is one insignificant cog in an enormously complex apparatus. He sees omniscience as something that can be achieved only by the sacred enterprise of Science with its uniquely privileged methodologies and brilliantly sophisticated instruments. He rarely gives thought to the possibility that every knowledge makes ethical demands on the knower. If he does so, he does so not as a scientist, but as an ethicist or a philosopher or a believer.

Traditional seekers of wisdom aim to actualize the full potential of intelligence in order to understand everything that is significant for human ends, and these ends are defined in terms of a metaphysics, a cosmology, a spiritual psychology, and an ethics that take Ultimate Reality as the measure of man. Modern seekers of facts aim to accumulate information and to devise ever more sophisticated theories in order to achieve what they call "progress." In other words, they want to achieve a transformation of the human race on the basis of scientistic and ideological pseudo-absolutes.

The quest for wisdom is qualitative, because it aims at the actualization of all the qualities present in the divine image and named by the names of God. The scientific quest for knowledge and theoretical prowess is quantitative, because it aims to understand and control an ever-proliferating multiplicity of things.

The more the traditional intellectual searches for omniscience, the more he finds the unity of his own soul and his own organismic relationship with the world. The more the modern scientist searches for data, the more he is pulled into dispersion and incoherence, despite his claims that overarching theories will one day explain everything.

The traditional quest for wisdom leads to integration, synthesis, and a global, anthropocosmic vision. The modern quest for information and control leads to mushrooming piles of facts and the proliferation of ever more specialized fields of learning. The net result of the modern quest is particularization, division, partition, separation, incoherence, mutual incomprehension, and chaos. No one knows the truth of this statement better than university professors, who are typically so narrowly specialized that they cannot explain their research to their own colleagues in their own departments – much less to colleagues in other departments.

As for the claim that science will soon achieve a theory of everything, this "everything" is in any case defined in mathematical and physical terms. Such a theory can have nothing to say about the higher levels of being, the first of which is the being of the knowing subject who declares himself the inventor or discoverer of the theory. By the necessities of its own presuppositions, science ignores that basic constituent of reality that is the very self of the scientist. Appleyard makes the point nicely:

> Scientific knowledge is fundamentally paradoxical. The paradox is that all of science's "truths" about the "real" world are based upon the most flagrant distortion. In creating an understandable universe, we have committed ourselves to the most gross and obvious oversimplification. We have excluded the understanding mechanism, the self.[26]

Moreover, the whole enterprise is built on the shifting sands of empirical observation and rooted in the imitation of the findings of others, not firsthand knowing. How can anyone know anything firsthand when all depends on observations made through scientific instruments and calculations by computers?

In short, for the Islamic intellectual tradition, the study of the universe was a two-pronged, holistic enterprise. In one respect its aim was to depict and describe the world of appearances. In another respect its goal was to grasp the innermost reality of both the appearances and the knower of the appearances. The great masters of the discipline always recognized that it is impossible to understand external objects without understanding the subject that understands. This meant that metaphysics, cosmology, and spiritual psychology were essential parts of the quest. The final goal was to see earthly appearances, intelligible principles, and the intelligent self in one integrated and simultaneous vision. It was understood that intelligence is not only that which grasps and comprehends the real nature of things, but also that which gives birth to things in the first place. Everything knowable is already latent within intelligence, because all things appear from intelligence in the cosmogonic process.

The anthropocosmic vision allowed for no real dichotomy between the subject that knows and the object known. The structure and goals of the enterprise precluded losing sight of the ontological links that bind the two. To do so would be to forget *tawḥīd* and to fall into the chaos of dispersion and egocentricity. Ignorance of the reality of the knower leads to the use of knowledge for achieving illusory ends, and ignorance of the reality of the known turns the world into things and objects that can be manipulated for goals cut off from any vision of true human nature.

The possibilities of human understanding define the possibilities of human becoming. To know is to be. To ignore the reality of either the object or the subject is to fall into foolishness, error, and superstition. An impoverished and flattened universe is the mirror image of an impoverished and flattened soul. The death of God is nothing but the stultification of the human intellect. Social and ecological catastrophe is the inevitable consequence of psychic and spiritual dissolution. Cosmos and soul are not two separate realities, but two sides of the same coin, a coin that was minted in the image of God.

7

The Search for Meaning

In the Islamic worldview, nothing could be more preposterous than to suggest that reality as we perceive it is just what it appears to be, or that human beings have at their disposal the means to plumb the ultimate depths of the universe and to solve all its mysteries. The basic given is that an inexhaustible richness of meaning and significance lies beneath the surface and beyond appearances. The Qur'an is full of verses that speak of the invisible realities that permeate the visible realm, realities that include God, angels, and spirits; indeed, the very foundation of Islam is "faith in the unseen" (cf. Qur'an 2:3). The primary unseen reality is God, who knows himself and, as the Qur'an tells us repeatedly, "all things." God alone, in Qur'anic terms, is "knower of both the Unseen and the Visible." As for human beings, "They encompass nothing of His knowledge save as He wills" (2:255).

Precisely because people are ignorant, they must search for knowledge. But this is not just any knowledge, nor is it information. Real knowledge takes as its object God and the doings of God (knowledge of reality as it is in itself), and God's guidance and instructions (knowledge of how human beings should act and be). Knowledge of both of these realms comes by way of the "signs" (*āyāt*), which the Qur'an locates in three broad domains: scripture and prophetic activity, natural phenomena, and the human self. The Qur'an's repeated use of this word announces that even though the significance of things and events is hidden, what we perceive gives hints and intimations of their meanings.

The fact that reality as we perceive it speaks to us of something far deeper and far more real follows directly upon *tawḥīd*, "There is nothing real but the Real." God alone truly deserves the epithet "reality," and everything else has an ambiguous status. God alone is Truth, and everything else simultaneously conceals and reveals the Truth.

TWO MODES OF KNOWING

The goal of the seeker of wisdom is to actualize and realize intelligence, which at its pinnacle is a transpersonal reality, fully aware of all of existence and dwelling at the very core of the human substance. Only by accessing intelligence can people find an eye adequate to wisdom, which is an attribute of God, the All-Knowing, the All-Wise.

Nowadays, something of the difference between the intellectual knowledge that aims for wisdom and the transmitted knowledge that depends upon imitation is reflected in the approaches and methodologies of the academic disciplines. Fields rooted in mathematics incline toward intellectual understanding, and fields having to do with history, social science, and the humanities are firmly grounded in transmission.

If mathematics was traditionally considered an intellectual science of sorts, this is because its principles can be discovered within oneself without the need for transmission. The special sense of certainty that comes from mathematical knowledge was seen as deriving from the fact that mathematics is an expression of the unitary, intelligible order that underlies apparent reality and forms the bedrock of the soul. Unlike transmitted knowledge, mathematical truths, once understood, are seen to be necessarily so, because they conform with the reality that shapes cosmos and soul. Nonetheless, to the degree that mathematics operates on the basis of data coming from outside the self, it was not considered a pure intellectual science. It partakes of a lesser degree of certainty and was commonly considered "intermediate" (*mutawassiṭ*) between transmitted and intellectual.

Most religious knowledge is transmitted, but most non-religious knowledge is also transmitted, because practically everything we know has been learned from others, not discovered within ourselves. Modern science inclines toward discovery, but what is discovered is typically thought to lie in the outside world, not in the inner world of the discovering self. Scientists would like to achieve firsthand knowing, but their general knowledge of science is necessarily transmitted. Given the *takthīr* that drives the accumulation of data and the proliferation of theories, scientists are by definition specialists, and even in their own fields, they discover nothing without building on the findings of their predecessors.

In short, modern science, especially its mathematical forms, has an "intellectual" proclivity, but at the same time, good scientists are the first to recognize that they stand on the shoulders of giants. To reach their goals, they take the received knowledge as given. It may happen that at a certain point, a large number of scientists question the transmitted theories to such a degree that they bring about a "paradigm shift." Then some of the authorities from whom they draw their transmitted knowledge and theoretical understanding will change.

True intellectual knowledge is altogether different. It is not achieved by standing on anyone's shoulders. Only what is known in the depths of the soul, without intermediary, is intellectual in the proper sense of the word. No one can pass such knowledge on to someone else, nor can it be found by reading and study. It must be realized within oneself through a long process of mental training and inner purification.

SUBJECT AND OBJECT

One of the fruits of intellectual learning was to understand – or rather, to see and realize – that the so-called "object" out there and the "subject" in here are essentially the same. To think of the two as separate is to falsify the meaning of cosmos and soul, to distort the relationship between things and self. Such falsifications inevitably lead to wrong relationships with self, people, and the world. The very structure of the intellectual quest stressed not only the achievement of right knowledge through the unification of subject and object, but also the actualization of sound moral character and the cultivation of virtue. The quest aimed at overcoming the soul's self-centeredness, to train it to detach itself from its individualistic tendencies, and to point the way toward bridging the gap between self and other.

Aspiring philosophers studied ethics as a standard part of their training, and Sufis considered the achievement of virtue and the avoidance of vice as the first priority. Ethics was not just a theoretical endeavor, but the guidebook for becoming a better person. At the same time, it was always taken for granted that correct activity – ethical, moral, and virtuous action – depends upon correct knowledge of the world, and correct knowledge of the world depends upon knowing the contingent and convergent reality of soul and cosmos.

How exactly the split between subject and object came to be firmly entrenched in the modern worldview has been much discussed and debated by historians and philosophers. Whatever the detailed reasons may have been, the result was that a separative, divisive epistemology gradually appeared and became crystallized with Descartes. For centuries, seekers of wisdom had understood that the highest purpose of knowledge was to achieve correct understanding of God and the world in tandem with self-understanding and self-realization. This approach was eventually abandoned and replaced almost entirely by another outlook. Knowledge came to be understood primarily as an instrument for control and manipulation. Certainly, many scientists remained ethical and moral human beings, but they could no longer address the necessity for virtue in the context of their own quest for knowledge of the natural world.

Originally, the search for wisdom went hand in hand with the attempt to perfect the soul. Philosophy, as Pierre Hadot has shown, had always been a way of life and a spiritual discipline.[27] Eventually, concern for the inner realm was relegated to theologians and moralists. Ethics was turned into an afterthought to "real" knowledge, and fact was disjoined from value. The premodern traditions had sought knowledge in order to cultivate and perfect the self, but the modern scientific enterprise abandoned the self to its own subjective realm and sought to manipulate and exploit the other. Few have explained what happened as well as Appleyard. In drawing a few conclusions, he remarks,

> Science trapped us all in our private reasons. It divided us from our world, locked us in the armored turrets of our consciousness. Outside was an alien landscape which was either illusory or meaningless, inside was the only possession of which we could be sure – the continued, anxious chattering of our self-awareness. Our souls were removed from our bodies.[28]

THE WORLDVIEW

The intellectual tradition held that the goal of study and learning was not to achieve a specific knowledge or to solve specific problems. Individualistic and specifying motivations were seen as diversions from the cultivation of the soul and the realization of selfhood. One could not love wisdom – which was none other than the Wise – by aiming to understand this or that, by attempting to achieve limited and defined goals. This is precisely what Avicenna means when he says, "Knowers desire the Real, the First, only for His sake, not for the sake of something else." Only accessing the intellect, the radiant light of the infinite God, allows for full actualization of the self and full understanding of the world.

What then is this "selfhood" that seekers of wisdom were striving to realize? This is the question I now need to address, with the caveat that true and real knowledge of selfhood is inaccessible to any but self. There is no object out there to be known. In knowledge of self, subject and object, knower and known, are the same thing. Moreover, any oral or written expression of self-knowledge can only be received by way of transmission. The only locus of intellectual knowledge is the knowing self. Transmitted expressions can at best point the soul in the right direction.

To suggest the nature of the self, we need a context in which discussing it makes sense. This is precisely the role of a worldview. The modern-day outlook on things – whether or not we accept the common idea that it is collapsing – does not provide an overview of the whole of reality, since real knowledge has been reduced to what can be verified empirically. Such verification, however, depends upon establishing some control over the object, a control that can only be obtained when subject and object are seen as distinct. Only external, controllable realms of reality are considered real, which is to say that "reality" has been reduced to the visible realm. The infinitely vaster realm of the Unseen is simply ignored.

Traditional worldviews are marked by a grandeur of scope that puts the invisible dimensions of reality at center stage. Their cosmological schemes have either been open-ended or could easily be understood as such by those who appreciate the language of symbolism and signs (contra the opinion of those who see the medieval, Christian universe as "closed"). In traditional worldviews, there are no limiting horizons, because any depiction of things must be recognized as a visible and inadequate representation of the Invisible. Phenomena are not opaque; rather, they are transparent, because they point to the Infinite – whether it be called God, Brahman, the Buddha-nature, or Tao.

From the standpoint of the intellectual tradition, the intuition of *tawḥīd* drives every quest for knowledge. All seekers of knowledge already understand at some level of their being that things are coherent, intelligible, and interconnected. Any healthy mind knows that the universe is held together by a single reality – the very word "universe" points to this intuition (even discussion of a "multiverse" is rooted in the unifying vision of human intelligence). The modern scientific enterprise illustrates the omnipresent intuition of *tawḥīd*, because it is built on the assumption that knowable laws govern the universe. Any talk of laws and knowability presupposes the notion of interconnection, interrelatedness, and ultimate wholeness. If some scientists choose to deny ultimate unity, they do so because it cannot be proven empirically, but their endeavors belie their words.

For the intellectual tradition, *tawḥīd* provides the only sure and certain point of reference, precisely because it announces the reality of the Absolutely One, the only reality that is truly real. Knowledge of the cosmos can then be derived by observing cosmos and soul while recognizing God as First and Last, Alpha and

Omega. The most typical word used to designate the Absolutely One in Islamic philosophy is *wujūd*, which, as we have seen, means not only being, but also finding, perception, awareness, consciousness, knowledge, joy. Consciousness is an essential attribute of the Real Being, which is to say that Being and Consciousness are exactly the same in the Ultimate Reality. It is this Being-cum-Consciousness that brings forth the phenomenal universe – that is, creates the world – by means of various attributes that are self-evident in our experience of ourselves and the universe, such as life, power, and love.

The Hindus tell us that Brahman is *sat–chit–ananda*, "being–knowledge–bliss." Seyyed Hossein Nasr has remarked that we can see an equivalent of this Sanskrit expression in the three Arabic words *wujūd–wijdān–wajd*, "being–consciousness–ecstasy," all of which derive from the same root *w.j.d.*, though each word stresses a different implication of Ultimate Reality.

To say, as the philosophers do, that God is "the Necessary Being" (*wājib al-wujūd*) means that by his very essence he is and cannot not be, but it also means that he is conscious and aware and cannot not be so, and that he is blissful and joyful and cannot not be so. These three attributes – being, awareness, bliss – then give rise to all the existential qualities that cause the world to coagulate out of nothingness.

The first reality that the Supreme Reality brings into existence, the Intellect or Spirit, is as similar to that Reality as any contingent thing can be. It is aware with a contingent awareness of all that may possibly be. The Real gives rise to multiplicity by means of this first, contingent reality. But the universe appears gradually and, as it moves further from its origin, becomes ever more diminished, just as the intensity of light decreases in keeping with its distance from its source.

This diminution of reality occurs in a series of stages that are enumerated in a variety of ways. The basic understanding is that the cosmos is coherent, ordered, layered, and directional. There are degrees of reality, some closer to Real Being and some further away. Closeness to the Real is judged in terms of the degree of participation in its attributes, that is, by the intensity of a level's unity, life, consciousness, power, will, compassion, wisdom, love, and so on. Distance from the Real is judged by the weakness of these same attributes. Ultimately, the traces of Being–Consciousness–Bliss become so attenuated that the process can go no further, so it turns back upon itself.

Muslim cosmologists see the universe as bi-directional, eternally coming forth from the Real and eternally receding back into the Real. It is at once centrifugal and centripetal. The Real is Absolute, Infinite, and Unchanging, and everything else is moving, altering, and transmuting. All movement is either toward the Real or away from it. The direction of movement is judged in terms of the increasing or decreasing intensity of the signs and traces of the Real that appear in things.

In this universe that is forever coming and going, there is no place for the stark dualisms that characterize so much of modern thought. In the more sophisticated cosmologies, reality is understood in terms of continuums, spectrums, complementarities, equilibriums, balances, and unities. Spirit and body, heaven and earth, past and future, local and non-local – all are understood as relative and complementary terms. Moreover, whenever a duality is discussed, there is typically a third factor, intermediate between the two, which plays the role of an "isthmus" (*barzakh*), something that is neither the one nor the other but allows for interrelationship. There was no terminology to express the stark dichotomies that Western thought has seen between "natural and supernatural" or "mind and body" or "spirit and matter." Everything natural has supernatural dimensions, and everything bodily is permeated with spirit; on every level the universe is infused with signs and intimations of unseen things. There can be no absolutes in any realm of observation – the only absolute is God, the One, who is Unseen and Unobserved by definition.

THE SELF

From the perspective of the philosophical tradition, the deepest root of the human self is the First Intellect, which knows every potentiality of phenomenal existence. It is this Intellect that gives birth to the universe in a centrifugal process analogous to the diffusion of light. As for the simultaneous centripetal movement, it appears wherever we look, especially in plants and animals, both of which show forth life and awareness.

Life, it needs to be remembered, actualizes a more intense degree of reality than lack of life. Life is an attribute of the Real, and among its traces are coherence and integration. In contrast, lack of life pertains to relative dispersion and incoherence. Moreover, life does not exist on the same plane as dead, inert, material things. We cannot analyze life per se, only its activities, signs, and traces. Life is already, in a profound sense, unseen and spiritual. Because life escapes fixity, it is less amenable than bodily things to mathematical analysis and technological manipulation. Its essential invisibility helps explain why biology can never be a "hard" science and why medicine will always be faced with the problem of determining the moment of "death."

Our only real knowledge of life is firsthand, inside ourselves. But where exactly do we know life? Life is essentially invisible and non-localizable, and this is even more true of awareness, which embraces the reality of life but simultaneously pertains to a higher level of being, further removed from inanimateness and closer to the First Real. Animal awareness, however, has severe constraints that become apparent as soon as we meditate upon the differences between human and animal possibilities. In effect, animals cannot transcend non-reflexive awareness of their environment. In contrast, human beings have the potential of moving beyond the limitations and constraints of the animal plane and of reflecting on the self that knows.

In other words, human beings can aim for "freedom" from their environmental limitations – not just physical limitations, but also social, political, and psychological limitations. Much more profoundly, they can strive for freedom from all limitations and all constraints. To do so they need to extirpate what the Buddhists call "the three poisons" – anger, greed, and ignorance. The basic impediments to freedom are the imperfections of the self, its failure to actualize its own reality. Ultimately, as Hindus well know, "freedom" (*moksha*) is the name of true and realized human selfhood. This is precisely realization, which is achieved by "freeing" or "disengaging" (*tajarrud*) the self from everything less than itself.

Inanimate things, plants, and animals are limited and therefore definable. Human beings are definable only inasmuch as they live beneath themselves. Any definition of human nature pertains to a level of being that lies beneath true selfhood. Definition pertains to realms that are essentially limited, such as the inanimate, the vegetal, the physiological, the animate, and the psychic.

Consciousness is not essentially limited; it itself is the subject that perceives limits, boundaries, and definitions. Strictly human modalities of being pertain to pure consciousness and pure awareness; the true human selfhood cannot be defined, yet it gives rise to every distinction and differentiation. Those human beings who fully realize their own selfhood – their innate, unlimited intelligence and consciousness – thereby gain freedom from every constraint.

It can also be said that there is no definition of the human self adequate to taking control of it and putting it to use. The self, in itself, is always free, despite the external constraints and controls that may be placed upon the bodily and animal planes of human nature, and despite the internal ignorance and illusion that typically veil the self from seeing things as they are and knowing its own freedom. Epistemologically, this means that true human selfhood cannot be the object of transmitted knowledge. It can only be known by direct, unmediated knowledge. We cannot know ourselves by reading about ourselves, carrying out controlled experiments, listening to what other people have said about us, or examining what we perceive of other people's selves. We can only know ourselves inside ourselves and without the intermediary of any instruments. These "instruments" include not simply scientific devices, but also the five senses, imagination, and thought, all of which are tools of the self.

In short, the human self per se dwells in a realm of being that transcends its own instruments. With even more reason, the Source of the self, which is the First Real, is inaccessible to the instruments of the self and even to the self itself. As the Sufis put it, "None knows God but God." Any real knowledge of God is simply the omnipresent God knowing himself through the human self, which is ultimately the First Intellect, the radiance of the Divine Light.

MEANING

Let me now turn to the question implied by the title of this chapter: how does one search for meaning in the intellectual tradition? It needs to be stressed that "meaning" is found by the knowing self inside itself, not outside. There is no "meaning" out there, over and apart from the observer. It is absurd to suppose that anything in the world can have a meaning apart from a self that is observing and understanding.

The connection between observer and observed goes back to the rootedness of all reality in the One Reality, which is Being–Consciousness–Bliss. We can understand this as signifying that God is object (Being), subject (Consciousness), and the living union of subject and object (Bliss) at one and the same time. In the universe, we initially perceive these three aspects of the One as distinct. The goal is to see all things as they truly are, and this demands reuniting the three aspects.

In the universe as we normally perceive it, subject and object are disjoined. The fact is, however, that the universe as object independent from a subject is not even there. I do not mean to suggest that the universe is contingent upon us as observers; rather, it is contingent upon the Necessary Being, the Real Knower, Brahman/Atman. The very being of the universe derives from the qualities and characteristics of the Real, whose traces it displays. The universe exists only as a "sign" of the Real, who knows it, perceives it, and understands it at every stage of its unfolding. In the last analysis, the universe has no existence save as an epiphenomenon of God's knowledge and consciousness. As some Sufis put it, the universe is God's dream, and as the Vedantists say, all is Maya.

The question of the search for meaning then comes down to this: can we know the meaning of the universe or of any object within it without knowing the meaning of the Real? Can we know the meaning of the dream without knowing the Dreaming Subject? Can we know our own selves anywhere else than within ourselves? Certainly, we can know the meaning of some things in relation to other things – all our disciplines provide this sort of meaning, though of course they provide it to those who understand the meaning, who find it within themselves. But what about the meaning of things as they really are – not in relation to other things, to this observer or that observer, but to the Absolute Observer, who is Being–Consciousness–Bliss? What about the meaning of things as they are situated in the infinitely complex web of intersecting journeys coming from the Real and returning to the Real? What in fact is the meaning of the individual jewels that stud Indra's net?

Given that human selves cannot be defined, they have no fixed standpoint within themselves. They have the potentiality of defining and understanding everything beneath their own level, and they have the ability to choose their standpoint in trying to understand. This means that people can look at the universe and themselves from a vast diversity of perspectives. The historical proliferation of cultures and worldviews is more than enough to show that the possible viewpoints allowing human beings to address the world and to search for meaning are beyond count. The proliferation of viewpoints, however, shows that the viewers are not in fact constrained in any essential way or confined to any specific viewpoint. Hence it is possible to step outside all viewpoints, all the ways of looking at the world that are conditioned by history, culture, religion, and science.

The great spiritual and contemplative traditions – traditions that are "intellectual" in the way I am using the word – are unanimous in declaring that it is indeed possible to become free of limitations and to act as the vehicle through which the Unobserved Observer observes. Human possibility transcends time, space, history, physicality, energy, ideation, the angels, and the gods themselves (though not "God" in the proper sense of the word). It is precisely this possibility of transcendence that marks the highest human calling. Indeed, when a tradition acknowledges this calling, it also acknowledges that this alone is the truly *human* calling. Every other calling turns people away from their root selfhood, which is the image of the Supreme Reality, if not that Reality itself. Every other calling represents misdirected love.

In short, the intellectual tradition maintains that the human self has the potential to go beyond every standpoint and every perspective, to step outside culture, history, and even the universe. The tradition sometimes calls the selfhood that achieves this freedom from all constraint "the standpoint of no standpoint" (*maqām lā maqām*), or "the Point at the Center of the Circle of Being–Consciousness–Bliss" (*nuqṭa wasaṭ dā'irat al-wujūd*). This ultimate standpoint is nonspecific and indefinable, so it encompasses every specific and definable standpoint. But, in order to reach the standpoint of no standpoint, one must harness the various dimensions that make up the external manifestation of the self's reality – body, soul, mind, thought, imagination – and attach them to the centripetal movement going back to the Center.

Despite these two movements – centrifugal and centripetal, descending and ascending – intelligence per se never leaves its own invisible and transcendent reality. In its deepest nature, the human self is indistinguishable from intelligence, so it remains indefinable and nonspecific. Every specific thing and every specific viewpoint tells the self what it is not. The self knows that it is not limited by the objects of its knowledge or by the finiteness of things, nor by the limitations of this standpoint or that science; it also knows that it has the potential to perceive and comprehend all definitions and all limitations. Hence it knows – if it is aware of itself – that it has no inherent limitations. It is free, not of this or that, but of all things, of everything less than the Real.

Reattaching oneself to the First Intelligence is the goal of aspiring "intellectuals." They want to make actual what is potential within themselves. But in order to achieve full realization, they must abandon dependence upon transmitted knowledge and come to know for themselves. To the degree that they do so, they rejoin the intelligence from which the soul departed at the outset and they achieve omniscience, though not in a differentiated way. This is a unitary understanding, an awareness of all things at their root. It is a spontaneous knowing, a blossoming of consciousness, an awakening to reality – all without reflection or thought. It is to see things as they are seen by the First Intellect before their appearance as coagulations in the universe.

From the standpoint of the intellectual tradition, every search for meaning that takes a specific standpoint – physics, medicine, sociology, theology – is constrained and limited by its premises and presuppositions. The discovered meaning will always be defined by the starting point. In contrast, in a purely intellectual quest, the only presupposition is the unity of the Infinite, Absolute, and Unknown Reality, which has no specific definition and stands in no standpoint. It is this non-specific goal that is sought by the seeker. The quest can have no closure, because the Infinite and Absolute can never be reached, though it reaches everywhere. As long as human beings take finite things, or a defined and known God, as the object of their quest, they can never know the true and final meaning of the universe and themselves.

Conclusions are inextricably linked to premises. Only the premise of *tawḥīd* – the transcendence, infinity, and absoluteness of the One Reality – allows the achievement of the full potential of the self. The conclusion of the quest will be the same as the first step, for no real steps can be taken without already being aware of the goal. At the beginning, however, *tawḥīd* is simply an inchoate intuition. It is then awakened and articulated by transmitted knowledge. Gradually it can grow into an actualized understanding, then a rational certainty, then a supra-rational comprehension of the way things are, and then a vision that transcends the vision of the eyes just as ocular vision transcends blindness. All these, however, are preliminary stages of consciousness. The goal is to realize *tawḥīd* for oneself and in oneself. One must find oneself and all things in their total context. The soul must come to recognize itself as a ray of the absolute and infinite Light. The beginning, then, is intuition and innate perception, and the end is the realization of being, knowledge, and bliss.

From the standpoint of this tradition, any search for the meaning of things and objects that does not allow seekers to open themselves up to the depths of their own selves will be an obstacle in the task of learning how to be human. It is impossible to know the meaning of anything without establishing a standpoint from which to speak of meaning. As long as the standpoint is determined by transmitted knowledge or theoretical frameworks, it will be limited by its givens. Only a standpoint of no standpoint can allow for transcending standpoints and arriving at the meaning behind all relative and situational meanings. The standpoint of no standpoint is available only in the transcendent realm that gives rise to the universe in the first place. True meaning can never be grasped by dogma, doctrine, theories, theorems, or any other mental construct. It can only be found by going beyond the operations of the mind, actualizing the unitary awareness of primordial intelligence that lies beneath the mind and behind the world, and integrating the human self back into its transcendent Origin.

Notes

1. *Cultural Schizophrenia: Islamic Societies Confronting the West*, translated by John Howe (London: Saqi Books, 1992).
2. For the Arabic text and another translation, see Michael E. Marmura, *Avicenna. The Metaphysics of the Healing* (Provo: Brigham Young University Press, 2005), p. 350.
3. The Annual Iqbal Memorial Lecture, Department of Philosophy, University of the Punjab, Lahore (November 10, 2000).
4. *Mathnawī*, edited by R. A. Nicholson (London: Luzac, 1925–40), Book 2, verses 277–9.
5. *Mathnawī*, Book 1, vs. 1406–7.
6. *Understanding the Present: Science and the Soul of Modern Man* (New York: Anchor Books, 1992), pp. 190–91.
7. In a treatise translated by Chittick, *The Heart of Islamic Philosophy* (Oxford: Oxford University Press, 2001), p. 182.
8. *The Reconstruction of Religious Thought in Islam* (Lahore: Iqbal Academy, 1986), pp. 33–34.
9. *Mathnawī*, Book 5, vs. 1289–93.
10. A great deal of scholarship traces the modern origins of fundamentalism. Karen Armstrong sums it up nicely in *The Battle for God* (New York: Ballantine Books, 2001), a detailed study of the history of the Jewish, Christian, and Muslim versions.
11. *Kīmiyā-yi sa'ādat*, edited by H. Khadīw-jam (Tehran: Jībī, 1354/1975), pp. 36–37.

12. *Rethinking Islam*, translated by Robert D. Lee (Boulder: Westview Press, 1994), p. 13.
13. *Mathnawī*, Book 1, vs. 1234ff.
14. *Understanding the Present*, p. 11.
15. *The Rise of Early Modern Science: Islam, China, and the West* (Cambridge: Cambridge University Press, 1993), p. 65.
16. *Religion and the Order of Nature* (New York: Oxford University Press, 1996), p. 287.
17. Tu in turn takes the word "anthropocosmic" from Mircea Eliade. Tu, *Centrality and Commonality: An Essay on Confucian Religiousness* (Albany: State University of New York Press, 1989), p. 126.
18. *Centrality and Commonality*, p. 10.
19. *Confucian Thought: Selfhood as Creative Transformation* (Albany: State University of New York Press, 1985), p. 137.
20. *Centrality and Commonality*, p. 102.
21. *Confucian Thought*, pp. 46–47.
22. *Understanding the Present*, p. 54.
23. *Confucian Thought*, p. 44.
24. *Al-Ishārāt wa'l-tanbīhāt*, edited by S. Dunyā (Cairo: 'Īsā al-Bābī al-Ḥalabī, 1947), vol. 3, p. 227.
25. *Confucian Thought*, p. 64.
26. *Understanding the Present*, p. 196.
27. *Philosophy as a Way of Life* (Oxford: Blackwell, 1995).
28. *Understanding the Present*, pp. 56–57.

Index

Absolute, absoluteness 65, 72, 74, 78, 101–2, 141, 145, 148
abstraction 15, 36, 81, 86, 87
Adam, as Muslim 112
  and the names 29, 30, 42, 84, 99, 128
  fall of 43, 112–13, 125
  see also human beings
afterlife 45, 66
agnosticism 72, 73
akhlāq 50
  ḥamīda 55
ʿAlī (ibn Abī Ṭālib) 27
angels 42, 76, 103, 105, 122
animals 85, 91, 121, 122, 124, 142–3
anthropocentrism 109
anthropocosmism 109–10, 127–8, 131
anthropomorphism 71, 87–8, 99
apologists, Muslim x, 95
Appleyard, Bryan 45, 86, 119, 130, 137
ʿaql 27, 41, 71
  bi'l-fiʿl 27
  bi'l-quwwa 127
  hayūlānī 127
ʿaqlī viii
Aristotle 120, 128
Arkoun, Muhammad 68–9
Armstrong, Karen 151
ascent 45, 89
atheism 72, 73
Atman 145
attributes (*ṣifāt*), *see* names
authority, and transmitted knowledge viii, 2, 3, 14–15, 24, 60, 65, 112, 113, 115, 118, 135
  new forms of 64
Avicenna (Ibn Sīnā) 28, 123, 138
awakening 2, 32, 56, 126, 128, 148
awareness, *see* consciousness
āyāt ix, 134

barzakh 141
bāṭil 120
bāṭin 81
Being (*wujūd*) 26, 92–3, 108
  Consciousness–Bliss 141, 145, 147, 149
  Necessary 140, 145
  Real 140–1
belief 16, 25, 48, 63, 64
biology 142
Bīrūnī, al- 94
Brahman 140, 145
Buddhism 104
bureaucracy 17, 56

certainty 35, 115, 135, 148
character traits (*akhlāq*) 50, 55
  see also ethics
Christianity 110–11, 116, 139
  decline of 118

Confucianism 109, 114
consciousness 28, 31, 41, 90–4, 115, 125, 143, 148
as *wujūd* 140
consensus 24–5
of scientific ulama 15, 25, 35, 64
Coomaraswamy, Ananda 78
Corbin, Henry 81
corruption (*fasād*) 18
cosmogenesis 32, 125
cosmology 26, 45, 82–3, 84–7, 92, 101, 131–2
scientific 86
cosmos (*'ālam*) 31, 91–2
birth of from consciousness 41, 125, 130
hierarchy of 31, 91, 105, 125, 140
purpose of 52, 87, 89, 123, 124
and soul 51–2, 77, 82–3, 89, 96, 101, 109–10, 114–15, 117, 122–3, 125–8, 131–2
two movements of 31–2, 141, 147
worlds of 31, 41
creation 11, 14, 42, 53, 84, 93, 104, 106, 120, 125
creationism 100
creativity 64

deiformity (*ta'alluh*) 50
Descartes 51, 137
dhikr 88, 113
dīn al-'ajā'iz 12
dogmatism 62–5, 67, 73–4, 116
dualism 141

ecology 83, 131
egocentricity 131, 136
empirical 24, 49, 56, 69, 83, 104, 119, 130, 138–9
engineering 77, 90
social 99
Enlightenment 51, 59, 65, 69, 109
environment, natural 18, 30, 34, 55
eschatology 32, 45
esoterism 81–2
ethics 17, 26, 65, 98, 106, 124, 126, 129, 136–7
objectivity of 54–6
evil 106–7
ḥaqq of 120
existence 90–1
experts 15, 20, 35, 36, 60, 86

faith (*īmān*) 4, 10, 11, 16, 110
objects of 16, 136
three principles of 40, 67
fatwa (*fatwā*) 15, 19
fiqh 3
fiṭra 29, 31, 33, 41, 47, 61, 112, 128
forgetfulness 29, 33, 42–3, 112–13
form (*ṣūra*) 122
Form-giver 122
freedom 32, 93, 120, 143–4, 147
fundamentalism 59, 73, 79, 80, 90, 96, 99
scientific 73

ghayb 42
ghayr ālī 61
Ghazālī, al- 46, 79, 80, 98
God/god 7, 12–13, 89–90
death of 62
love for 123, 138
manifestations of 90, 91, 93, 125
names (and attributes) of 7, 50, 52–3, 54, 70–1, 84, 88–92, 93, 96, 98, 102–3, 105, 106, 140
transcendence (and immanence) of 71, 72, 83, 87–8, 91–3, 95, 101–3
varieties of 12–21, 62, 67–8
government 21, 56, 101
Grunebaum, Gustaf von 75
guidance (prophetic) 78–9, 85–6

Hadith 2, 4
hadiths cited: 7, 12, 21, 27, 75, 76, 77, 89, 90, 120
Hadot, Pierre 137
ḥaqq 45, 50, 55, 63–5, 69, 71, 72, 119–20, 126, 127
God as 43, 63, 93, 119–20
hawā' 55
ḥayawān nāṭiq 85
hearsay viii, x, 2, 6, 24, 25, 66, 68, 119
heart (*qalb*) 26, 27, 31, 71
ḥikma 49

Hinduism 88, 104, 140, 143
historicity of transmitted knowledge 110–14
history, *see* intellectual, thought, Western
Huff, Toby E. 97–8
human beings (*insān*) 31, 41
  becoming human 110, 126, 128, 149
  cosmic role of 52–5, 75–6, 85–6, 97–8, 103–5, 122–3
  diversity of 32, 76, 124
  divine image of 31, 51, 82, 87, 105, 113, 123, 136, 131
  goal of 43–4, 53, 109, 126, 148
  indefinability of 142–3, 146; *see also* self
  innate disposition of 29–30, 33, 41, 42, 111–12
  ontological root of 41, 53
  speech of 85–6
  see also perfection, self
ḥuqūq 65, 71, 93
hylomorphism 122

ʿibāda 112, 123
Ibn al-Haytham 94
Ibn ʿArabī 70–3, 88, 95
Ibn Ṭufayl 33
ideology 1, 59, 73–4, 99, 124, 129
  and imitation 64
  in modern Islam 40, 67–9, 82, 89, 90
ignorance 29, 94
  compound 20, 30
ijmāʿ 15, 23
ijtihād 3–5, 45
ʿilm 78
imagination 61, 88
  modern 70
  mythic 69–73, 88
  and reason 70–2, 88, 95
īmān 4
imitation (*taqlīd*) 2, 3–5, 56, 63–4, 119, 134–5
  Muslim criticism of 28, 45–6, 63–4
  scientific 15, 34–5, 56, 119, 130
  wrong-headed 15, 18–21, 36, 47, 68, 74
incarnation 111
Indra's net 146
information 13, 26, 34, 133–4
intellect, intelligence (*ʿaql*) 27–33, 41–2, 46, 61, 65, 98, 104, 125–8, 147
  actual 27–8, 125, 127–8, 134, 138, 148
  First 142, 144, 148
  practical and theoretical 30, 33
  purpose of 5, 124
  root of 32
  transpersonal 55, 73, 127, 130, 134
  Universal 53
intellectual tradition, traits of 8–10, 18–20, 34–5, 44, 55–6, 110, 128–32
  goal/role of 5–8, 25–9, 30, 44–5, 47–50, 55–7, 59, 61, 67–8, 131–2
  history of 8, 9–10, 48–50, 56, 62, 94–5, 116–18
  topics of 1–2, 16, 28–9, 30–3, 44–5, 131
  see knowledge
Iqbal, Allama 39, 45, 54
islām 111
Islam 6, 7, 8, 111
  in modern times 8–12, 17–20, 40, 89, 100–1
Islamism 68, 101
  see fundamentalism

jabbār 14
jahl murakkab 20
jurisprudence, jurists 3, 45–6, 71, 81, 88, 114
  dogmatism of 60, 62, 95
  modern-style 63–4, 70

Kalam, *see* theology
kalimat al-tawḥīd 7
Kant 51
Kāshānī, Afḍal al-Dīn 53
kashf 25
khalīfa 42
khayāl 71
knowledge, goal of 34, 109–10, 137
  infinity of 127
  intellectual vs. transmitted viii–x, 1–8, 23–9, 43, 45–6, 60–5, 65–9, 110–21, 135, 138, 144, 147–8
  non-instrumental 61, 135, 144
  quest for 34, 128–30, 134–6, 138–9
  self-knowledge, *see* self

knowledge (*cont.*):
sensualist 83, 104
unity of knower and known in 73, 131

law, *see* jurisprudence, Shariah
life 121–2, 142
divine 90–1, 92
light 142
of God 31, 42, 50, 96, 125, 144, 148
love 43, 55, 91, 104–5, 141
for God 123, 138

ma'ād 31, 44
mabda' wa'l-ma'ād, al- 31, 44
mādda 122
Man (*rajul*), of Number 76, 102
of Unseen 75–7, 101–8
maqām lā maqām 147
Marxism 112
mathematics 2–3, 5, 86, 94, 102, 134–5
matter (*mādda*) 122, 141
Maya 145
meaning ix, 12–13, 36, 68, 87, 89, 93, 97, 133–4, 144–9
loss of 17, 37–8, 89, 137
mercy (*raḥma*), precedence of 88–90, 92, 107
metaphysics 26, 45, 94
death of 62
of science 97–8, 104, 112
microcosm and macrocosm 122–3, 124, 128
modernism, Muslim 80, 89, 96, 99, 107
modernity, characteristics of 12–16, 48, 62, 70, 87
coerciveness of 99, 101
false thinking of 40, 49, 52–5, 69–70
gods of 12–16
prophets of 19, 65, 112
see also Western tradition
modernization 17
moksha 143
morality, *see* ethics
Muhammad 66, 111, 113
muḥaqqiq 4, 25, 46, 57
mujtahid, see *ijtihād*
mukaththir 53
Mullā Ṣadrā 44
multiplicity 91, 103, 140
muṣawwir 122
mutawassiṭ 135
muwaḥḥid 53
myth 69–70, 72

nafs 27, 121
najāt 127
names, divinely taught 29–30, 31, 85, 90, 98, 105, 128
see God
naming 84–90, 105–6
efficacy of 84, 96–8
schemes of 97
naqlī viii
Nasr, Seyyed Hossein 75, 77–83, 94, 99, 107, 140
nāṭiq 85
nature 72, 100, 107–8
study of 123–4
New Ageism 79
Newton 95
nuqṭa wasaṭ dā'irat al-wujūd 147

object, *see* subject
objectivity 54, 93–4, 99, 105, 119
omniscience 128–32, 147
Orientalism 77, 80
origin (*mabda'*) and return 7, 32, 44–5, 53, 125–8, 140, 145, 149
originality 19
orthodoxy 95
scientific 100

perennialism 78
perfection (*kamāl*) 27, 28, 33, 45, 49–52, 66–7, 76, 127
philosophy ix, 28–9, 31–2, 41, 44–5, 100, 114–18
criticism of 66, 67
history of 44, 66–7, 94
and prophecy 66
physicality 104, 105
Poerksen, Uwe 14
Pole (*quṭb*) 76–7, 102
politicization (of religion) 18, 59, 101
polynomiality 88, 102

polytheism 35, 36, 88
see *takthīr*
prayer 88, 99–100, 101
priesthood, of science 36
principles (of faith), three 40, 67
progress 14, 15, 17, 20, 35, 62, 68, 129
scientific 68, 94, 95
prophecy, prophets 40, 76, 79, 85, 112
of modernity 19, 65, 112
as source of transmitted knowledge 66–8, 112, 113
universality of 41, 79, 97–8, 113
purpose, of existence 52, 87, 89, 114–15, 123, 124
psychology, spiritual 26–7, 44, 53, 82, 84, 94

qadīm 19
qalb 27, 71
qaws, nuzūlī 32
ṣu'ūdī 32
quantification 96, 99
and quality 96, 100
Qur'an 2, 4, 18, 111, 113
verses cited: (2:3) 133; (2:31) 29; (2:255) 133; (2:286) 6; (3:83) 111; (7:156) 89; (9:67) 43; (10:47) 112; (11:7) 100; (13:17) 100; (14:4) 112; (21:25) 112; (21:30) 100; (24:35) 96; (24:41) 99; (30:41) 18; (57:4) 88

rationality 85, 88, 95
instrumental 69, 70, 110
scientific 72, 96
reality, God as 43
realization (*taḥqīq*) 2–4, 23, 32, 46, 63, 143
goal/role of 23, 25–9, 46–7, 50–1, 56–7, 72, 135
methodology of 65, 85, 118–21
of self 115, 136–7, 147
reason (*'aql*), according to Ibn 'Arabī 72–3, 88
and revelation 110–11, 115–16
reification of cosmos 83, 101
religion 32, 60, 88
dismissal of 37–8, 48
politicized 18, 59, 101
remembrance (*dhikr*) 43, 88, 113
responsibility 46, 63, 65, 71, 83, 93
return (*ma'ād*), *see* eschatology, origin
revelation (scripture) 66, 71, 72, 89, 93, 112
see reason
rights (*ḥuqūq*), of God 73
human 63, 73
rijāl, al-'adad 76
al-ghayb 75
ritual 48, 88
Rūmī 42, 43, 44, 57, 85

sa'āda 127
saints 76
ṣalāḥ 18
salvation 127
sat-chit-ananda 140
Satan 106–7
Schuon, Frithjof 78, 81
science, scientists (modern) 11–12, 15, 19, 34, 35–6, 86, 97–8
and Islamic science ix, 44, 48–9, 56, 94–8, 104, 110, 115–16, 119, 128–32
and control 55–6, 95, 109–10, 129, 137, 138
and discovery 25, 56, 68, 135
as excluding human concerns 51, 54–5, 62, 83, 86–7, 92, 97, 107–8, 129–30, 137, 138–9
as illustrating *tawḥīd* 139
metaphysics of 97–8, 104, 112
as myth 70, 83, 90
prophetic/priestly role of 25, 36, 67
as *takthīr* 52, 77
as transmitted knowledge 26, 119, 130, 134
unintelligibility of 54, 86, 130
scientism 48–9, 68, 70, 77, 82, 83, 105, 117–19, 129
secularism 17, 62, 118
self, selfhood (*nafs*) 33, 55, 142–4
indefinability of 32, 138, 143–4, 147
knowledge of 34, 51, 56, 73, 130, 136, 145–7
see also soul
senses 60, 61, 90, 116
service, *see* worship

shahāda 42
Shahadah 7, 41, 97
two parts of 66, 111–13
shahwa 55
Shariah 3–4, 45–6, 77, 90, 96
Shayegan, Daryoush 11
Shi'ism 3
shirk 12, 47–8
ṣifāt 50
signs (*āyāt*) of God ix, 6, 11, 12, 69, 72, 134, 139, 140, 141
ṣirāṭ mustaqīm 7
soul (*nafs*) 27, 103, 121
degrees of 121–2
immortality of 66–7, 87
as potentiality 127, 147
study of 44–5, 61, 121–4
see also cosmos, self
spirit (*rūḥ*) 341, 43, 103, 125, 140
blown into Adam 30, 42, 72, 125
and body 41, 140
divine 50, 55, 61
standpoint of no standpoint 147, 149
subject and object 51, 122, 127, 131
unity of 109–10, 116, 123–4, 128, 136–7, 145
Sufism ix, 25, 32, 45, 76, 79–82, 92, 103, 136
criticism of 65, 79, 99
Sunnah 2, 6, 89, 96
Sunnism 3
superstition 37, 94, 96, 99–101, 131
ṣūra 122
symbolism 69–70, 72, 96, 115, 139

ta'alluh 50
ṭabī'a 122
taḥqīq 2–3, 23, 45–6, 50–1, 56–7, 63, 72, 85, 118–21
see realization
tajarrud 143
takhalluq 50
takthīr 12–14, 16–17, 19, 35–6, 52–3, 61, 72, 78, 135
as divine attribute 52–3, 90
ṭalab al-'ilm 20–1
taqlīd 2–3, 45–6, 63–4
see imitation
tawḥīd 7, 12–14, 16, 29, 30–1, 40, 46–7, 57, 61, 66, 67, 90, 148–9
ahistorical nature of 41, 66, 111–13
forgetting of 17, 18, 47, 116, 131
meaning of 43, 52, 73, 84, 101–2, 111–12, 120, 148
as orientation 52, 96, 139
as seed and fruit 50–1, 125
technology 17, 56, 70, 83, 99, 142
teleology 116
see purpose
theology (Kalam) 69, 71, 82, 85, 113–14
agreement of with philosophy 66–7, 88
dogmatism of 62, 63
new-style 63–4, 70
rationalism of 71–2, 82, 88
theomorphism 87, 97, 99
thought 41–2, 49, 55
modern Islamic 8–12
traditional Islamic 5–8, 12, 39–44, 47–8
see also intellectual tradition
totalitarianism 17, 101
tradition 18–21
traditionalism 78
transcendence, as human calling 146, 148
see God
transformation 50, 115
of human race 129
transmitted, *see* knowledge
trust 25, 60–1, 90
truth 46, 48, 63–4, 119, 130, 135
and ethics 54–5
of intellectual vs. transmitted knowledge viii, 24–9, 34–5, 41, 50, 57, 61, 64, 67–8, 73–4, 110–14
tawḥīd as first 41, 46, 112, 115
Tu Weiming 109, 114, 117, 121, 122, 127

ulama (*'ulamā'*) ix, 2, 15, 67–8
unity, *see tawḥīd*
universe, *see* cosmos
Unseen (*ghayb*), Men of 75–7, 101–8
and Visible (worlds) 42, 104, 105–6, 133, 138–9

value and fact 137
Vedantists 145
verification, *see* realization
vicegerent (*khalīfa*) 42, 53, 56, 106, 123
virtue (*faḍīla*) 27, 55, 136
Visible (*shahāda*), *see* Unseen

wajd 140
wajh 73
wājib al-wujūd 140
walī 76
Western tradition, history of 51–2, 71–2, 96, 110–11, 115, 137, 141
wijdān 140
wisdom 29, 34, 49–50, 54, 107, 125
  quest for 62, 114–18, 129, 134, 137
words, plastic 14–16
worlds (*'ālam*), two 31, 103–4
worldview 60, 86, 137, 147
  modern (scientific) 24, 35, 48, 54, 65, 77, 82–3, 86, 109
  traditional Islamic 11, 41, 54, 77–8, 85, 110, 133–4, 138–9
worship (*'ibāda*) 13–14, 18, 112, 123
wujūd 90–1, 92–3, 96, 104, 140

www.ingramcontent.com/pod-product-compliance
Lightning Source LLC
Chambersburg PA
CBHW030001110526
44587CB00011BA/951
* 9 7 8 1 8 5 1 6 8 4 9 5 3 *